Disaffection with School Mathematics

Disaffection with School Mathematics

Gareth Lewis
University of Leicester, UK

SENSE PUBLISHERS
ROTTERDAM/BOSTON/TAIPEI

A C.I.P. record for this book is available from the Library of Congress.

ISBN: 978-94-6300-329-2 (paperback)
ISBN: 978-94-6300-330-8 (hardback)
ISBN: 978-94-6300-331-5 (e-book)

Published by: Sense Publishers,
P.O. Box 21858,
3001 AW Rotterdam,
The Netherlands
https://www.sensepublishers.com/

Printed on acid-free paper

TABLE OF CONTENTS

PREFACE

The research I have been undertaking since 2009 is deeply rooted in my personal experience. I was a teacher of secondary mathematics, and, more latterly computer studies from 1977 to 1986 in working class and inner city schools in the UK. During that time I rose from a basic grade teacher to Head of Department.

Those years were interesting and exciting times for the teaching of mathematics. They were times of optimism and great innovation. Schools were changing with the demise of grammar schools, and the rise of comprehensivisation. But for me, the most significant changes were the changes in the school mathematics curriculum, and the teaching of mathematics. I taught at two schools, both of which had recently changed from grammar to comprehensive. In addition, the school leaving age had recently been raised to 16 years. In the school in which I taught, there was an urgent desire to find ways to teach maths to the 'lower ability' pupils. From the beginning, in both of these schools, I was assigned the role of 'young moderniser'. My role was to move away from the grammar school curriculum and outdated teaching methods, and I accepted the challenge with relish. In this way, from the start, I had a special affinity with the 'lower' groups, and with the newer programmes of study and teaching materials.

These exciting times saw a proliferation of initiatives that brought innovation and excitement to the whole enterprise of teaching mathematics. We had SMP (Schools Mathematics Project), SMILE cards, The Association of Teachers of Mathematics (ATM). I also joined the British Society for the Psychology of Learning Mathematics (BSPLM, now renamed BSRLM). I remember attending conferences and hearing RR Skemp, Bill Brookes and Dick Tahta speak. I was inspired by Starting Points by Banwell, Saunders and Tahta to make a set of workcards to use with my pupils. I eagerly devoured other books including everything by Skemp, Laurie Buxton's book on maths anxiety, John Holt's 'How Children Fail'. Later on, I became fascinated by the work of Papert and others – 'Mindstorms' became, and still is, one of my favourite reads. I obtained one of the early versions of Logo, which I used on our BBC model 'B' computers. I wrote my Masters dissertation on the use of computers in mathematics education in 1986.

Looking back on those years, a number of strong convictions developed and worked their way into my approach as a teacher. One was that mathematics did not have to be boring and a chore; on the contrary, it could be exciting, interesting and relevant to any pupil, at any level. Alongside this a more subtle conviction was also developing, and that was borne of frustration that pupils could leave school at 16 years old still not having mastered the basic operations that they were introduced to in primary school. It bothered me if those young people could not do maths, as it bothered many of them. It was quite clear to me, even then, that this lack of competence was not (as many would have it), a matter of ability. Although mathematics was used as a way of ordering pupils, I was convinced that those relegated to lower groups could 'master the basics' if they really wanted or needed to. But sadly, too many of them didn't. I knew then that issues of psychology were deeply embedded in this problem, without, perhaps, being able to articulate just exactly how.

Another book which influenced me was Children's Minds by Margaret Donaldson (1978). I found so much that resonated with my own views and experience at the time. I will limit myself to this quote, which gets to the heart of the matter:

> In the first few years at school all appears to go very well. The children seem eager, lively, happy… However, when we consider what has happened by the time children reach adolescence, we are forced to recognise that the promise of the early years frequently remains unfulfilled. Large numbers leave school with the bitter taste of defeat in them, not having mastered even moderately well those basic skills which society demands, much less having become people who rejoice in the exercise of creative intelligence. (p. 13)

My experience in education gave me a frustration about unfulfilled potential ('the bitter taste of defeat'), an affinity for student-centred approaches to learning, and curiosity and some knowledge about the transformative power of learning. The failure of so many young people to learn even the 'basics' of mathematics (some would argue it is numeracy, not mathematics at issue here) bothered me then, and it bothers me now. When, much later on, I had the opportunity to return to University to research mathematics education, this was the topic that I felt most passionate about. When I did this, I was able to understand that the psychological issue at hand was to do with motivation and affect.

This account of my research will begin with an examination of my understandings of the importance of affect (and therefore disaffection) in learning or not learning mathematics. From here, I will review literature and schools of thought on affect in mathematics education, and develop my argument and position that motivation and emotion can be seen as the central foundations of the affect space. This position leads to a brief description of the theory and methods that underlie the research, before presenting the results of the research. These results, in turn, are presented, in effect, in two halves: a series of case studies, and a more rounded consideration of the various and diverse aspects of the experience of disaffection.

ACKNOWLEDGEMENTS

It is impossible to embark on a doctoral study without acknowledging that the people to whom I am connected and who have influenced me, are an essential part of the journey. There are so many such people that it is not possible to name them all, but some deserve a special mention.

Undertaking this task was a life-changing decision and change of direction for me, and that cannot be done without love and support from family. So my thanks and appreciation go to Denise, Bryn, Joe and all of the Lewis's for putting up with me, indulging me in my eccentric choices and keeping me grounded.

It has been a complete joy to be the 'young' researcher at the University of Leicester where staff and colleagues have welcomed and supported me. In particular, I want to thank Janet for all she has done to get me to this stage. Most of all I would like to thank her for insisting on calling me 'young man' and allowing me to be a naïve and enthusiastic apprentice researcher.

For a researcher, data is the main currency, and access to schools, colleges and students is vital. I am aware that it is a gift, and I have been blessed in having had so much cooperation and goodwill from a number of people, who gave time and effort for little reward. Thanks go to Janet and Mel G. Special thanks go to Mel P for giving me such fantastic opportunities and help, and for making me feel welcomed as part of the team. Thanks also to Brenda for 'giving' me a whole school year to work with – the data is extremely precious.

Michael Apter has been a part of my life for many years now. He, along with Mitzi and others welcomed me into the Reversal Theory community, and he has been a colleague and friend during all of those years that we worked together. He encouraged me in undertaking this study, and has been incredibly generous with his time and support through this process.

It is not possible to overestimate how much Dr Neil Sellors has done to challenge, support and cajole me through this process. From the very first discussion in the pub with a more-or-less blank piece of paper, through the many, many sessions (also, coincidently in the pub) when we have discussed matters research, I couldn't have wished for or expected the commitment he has put in on my behalf. I appreciate all of the hours he has spent poring over first drafts and providing forensic feedback. He has helped me to think and write a little like a scholar, and one day I will be half as good as he is.

Finally, if I have resources and belief enough in my ability to take on this challenge, I owe it to Doff and Ivor Lewis. It gives me pleasure to think how proud you would have been, so this is also for you.

CHAPTER 1

THE PROBLEM WITH MATHEMATICS

> If I had to pick out a fundamental reason why this minority get turned off, I would have no hesitation: boredom...boredom is the bane of education...the recruiting sergeant for disaffection, truancy and bad behaviour.
> (David Miliband, Schools Standards Minister, in a speech to an education conference in London, February, 2003) (Accessed on October 27, 2010 on news.bbc.co.uk/1/hi/education/278i525.stm)

Disaffection with school mathematics is the primary concern of this book. However, the landscape of disaffection is wider than the study of school mathematics alone. The quote above is an interesting statement since it encapsulates a number of ideas that are the main concern of this study. It suggests that disaffection is high on the agenda for consideration by a politician with an interest in education. It shows further the conflation of the notions of disaffection with social issues (in this case truancy and bad behaviour), as well as issues of emotion and affect (boredom). It isn't necessary to agree with the politician to accept the premise that disaffection with education is seen as a concern that potentially has serious consequences.

THE IMPORTANCE OF DISAFFECTION

I will introduce the general issue of disaffection with education in order to paint a picture of the landscape within which specific disaffection with school mathematics sits.

Concern with disaffection with education is global. The Report Card of the OECD (Adamson, 2007) on child well-being, paints a gloomy picture, particularly for the UK. The UK has the lowest average rank (18.2) overall on the report card, with less than 20% of respondents agreeing with statements such as 'I like school a lot', which is less than half the proportion of nations like Norway, Austria and the Netherlands. The report paints a bleak picture of the ability of the UK system to educate and promote well-being through education. Article 29 of the UN Convention on the Rights of the Child talks of the importance of 'the development of the child's personality, talents and mental and physical abilities to their fullest potential' (Adamson, 2007, p. 19), and it is reasonable to argue that the UK is failing in this ambition.

In the UK considerable attention has been paid to the problems associated with low attainment in education, but the affective correlates of disaffection and disengagement have also been studied. Research undertaken by NFER (Kinder et al., 1996) has been influential in enhancing our understanding of pupils' views on the issue. Part of the rationale for this is an understanding that the disaffection and disengagement that is assumed to lie behind truancy, anti-social behaviour and exclusion, is known to be damaging to the life outcomes for the pupils involved. The UK policy think tank Demos (Birdwell et al., 2011), in a pamphlet 'The forgotten half', examines how well the UK educational system is serving the 50% of pupils who do not go on to university and higher education. Although its primary focus is employability, it concludes that pushing pupils through the hoops of assessment designed to lead to university may be at the root of the problem of disaffection and low attainment.

A number of conclusions can be drawn from this evidence. Firstly, there appears to be widespread concern about levels of attainment in the UK, and the ability of the school system to meet the needs of young people and of the wider society and economy. Secondly, there appears to be a strong connection made between levels of attainment and affective dimensions such as disengagement and disaffection.

Attempts have been made since the 1990's to apply research to an understanding of the landscape of disaffection and low attainment, and to try to understand the causes (Kinder et al., 1996). Noss (2009) focused on the low or underachievement of boys relative to girls in secondary schools in the UK and points out that disaffection and disengagement are primarily issues of motivation, since he emphasises intervention strategies that link to attitudes, expectations and aspirations. Noss found that the category 'disengaged' accounts for 12% of boys in year 9, rising to 20% at year 11. Again, the causal chain that begins with disengagement leading to truancy and poor or risky behaviour, which in turn leads to poor employment and life outcomes, is emphasised.

Lumby (2012) argues that "from a global perspective, the position of youth is calamitous' (p. 262). He makes a compelling case that the education system in the UK is failing young people, variously describing: "the evidence of dysfunction is compelling" (p. 275), and "Mainstream provision which is not appropriate for a significant proportion of learners" (p. 276). He characterises the issue of disaffection as an affective issue. In contrast to much of the literature cited above, Lumby is not driven by concerns of economics or employability, but by issues of equity. He cites Nussbaum approvingly, and introduces notions of capability and well-being. He points out that happiness or psychological health has entered the discourse of education, and is linked to notions of social justice, and this in turn emphasises that affect (and therefore disaffection) needs to be central in consideration of issues of learning and education.

In terms of competence, Lumby identifies the impoverished range of pedagogic practices that cause boredom and disengagement. He talks of the "unvarying didactic pedagogy" (p. 269) which does not meet the need of many pupils. Lumby's analysis thus provides some important ideas to inform the current study. The centrality of psychological and particularly motivational issues is a key idea. Methodologically, he also emphasises that young peoples' voice must be taken seriously, and he suggests that his aspiration is "authentic listening with sustained attentiveness"

(p. 267). There is also an interesting discussion here in that he notes that students had a limited vocabulary for describing aspects of their experience of education, and thus that care and novelty in research design are needed to understand more fully the emotional landscape of disaffected pupils.

DISAFFECTION WITH MATHEMATICS

The study reported here is concerned with the incidence and nature of disaffection with school mathematics, and it is not difficult to understand the wider concern for standards of attainment in school mathematics, which are seen as both a cause and a result of disaffection.

This concern about mathematics education in the UK is not new. I was a mathematics teacher in the 1970's and 1980's. Prime Minister James Callaghan's challenge to education in 1976 was followed by the Cockcroft Report (1982) which reflected concerns about mathematics education. In more recent years, there has been a raft of inquiries and reports relating to the provision of mathematics education and attainment in mathematics. The State of the Nation Report into Science and Mathematics Education by the Royal Society notes the widespread nature of current concern: "no decade since the 1970's...has seen so much being written about the disaffection young people appear to have for science and mathematics" (The Royal Society, 2008, p. 171). The strategic importance of mathematics nationally is demonstrated by the STEM (Science, technology, engineering, mathematics) initiative. In the public domain, concern is focussed primarily on standards of attainment. Organisations like the Confederation of British Industry regularly voice their concern over standards. In its report 'Making it all add up' (CBI, 2010) it states "At present, not enough young people leave school or college with the numeracy and maths skills they need for work and life" (p. 2). On its 'report card' it states pointedly "The UK must do better if it is to reach its full potential" (p. 3). The CBI's concern of course is with numeracy and a utilitarian agenda for business rather than with mathematics education in the round. But concern for mathematics attainment is shared by the wider public. In her own report, Vorderman (2011) talks about the corrosive effect of frequent failure, and the damage that this causes. The report espouses an approach to mathematics education that goes wider than the purely utilitarian. It talks about 'entitlement', and mentions not just achieving success, but also of 'satisfaction' and of 'increased confidence and motivation' (p. 22). This wider rationale for studying mathematics in schools is endorsed by The Royal Society who suggest one of the purposes of learning mathematics is:

> To enable as many students as possible to participate in the scientific and mathematical elements of the conversation of humankind, in as many settings as possible. (The Royal Society, 2008, p. 21)

Further evidence on the notion of the effect on young people's lives of disaffection with mathematics comes from the charity Every Child a Chance Trust, claiming that research shows that weak mathematics skills are linked with an array of poor life outcomes. In its report 'The long term cost of numeracy difficulties', they say:

> Competent numeracy would thus appear not only important in relation to employability and the economy, but also as a protective factor in maintaining social cohesion. (Gross, 2009, p. 4)

The report also points out evidence of links between poor numeracy and issues of health, antisocial behaviour, employment, crime and other negative life outcomes.

In the inquiry into the state of mathematics teaching and learning in the UK (Smith, 2004), the then secretary of state, in his forward says:

> The Inquiry has therefore found it deeply disturbing that so many important stakeholders believe there to be a crisis in the teaching and learning of mathematics in England. (p. v)

He goes on to talk about: "The failure of the curriculum to excite interest and provide appropriate motivation" (p. 4). There is a strong recognition in this evidence that unless we can understand disaffection, we won't be able to tackle the problems of attainment and progression in mathematics education, or indeed the social and economic problems that are so closely associated with low attainment. Understanding how and why disaffection occurs is the most important foundation for doing something about it.

In summary, there is a crisis in mathematics education in the UK, which is shared across many countries. This has been driven to a large extent by concern about levels of attainment, but supplemented by concern about participation and progression. It has been argued here that to learn mathematics not only for utilitarian purposes, but also to enjoy the experience of learning mathematics, and to be a fully rounded citizen, are also important ambitions.

I have adopted a position then, that a competent grasp of mathematics is a condition of citizenship and social functioning, since the consequences of competence are associated with the practical benefits of a qualitatively better life as a citizen. This discussion leads naturally to the question of whether the learning of mathematics and achievement of competence in mathematics is a human right. There is a strong implication of such an assumption in the work of Lumby (2012) and others. In the international arena, there seems to be widespread appreciation that education is about more than just utilitarian issues. The debate, more recently, has moved from discussion of rights to the notion of capabilities (Nussbaum, 1997). Central to the notion of capabilities is that people should be free to achieve 'functionings', which are states that make up peoples' well-being, and which are important to them. In this way, the notion of capabilities stress non-economic values. Education is seen, not just in utilitarian terms, but as intrinsically valuable for human flourishing.

> For this reason, I argue that the most appropriate space for comparisons is the space of capabilities. Instead of asking "How satisfied is person A," or "How

> much in the way of resources does A command," we ask the question: "What is A actually able to do and to be?" In other words, about a variety of functions that would seem to be of central importance to a human life, we ask: Is the person capable of this, or not? (Nussbaum, 1997, p. 285)

Nussbaum not only critiques the human capital approach to evaluating achievement and allocation of resources, but also the social-cultural approach that focus on the notion of power and access to resources.

The capabilities approach is of considerable interest in education, but the implications for education are at an early stage of being examined and explicated. A moderate but justified conclusion would be that, when evaluated in capability terms, mathematics education will be about much more than utilitarian issues, such as contribution to the economy or access to employment. In the sense discussed here, the notion of capabilities is seen as more helpful than the notion of rights. In addition, and relating the debate directly to the notion of disaffection, it is easy to see that disaffection can be viewed as a de-limiting factor in the development of capabilities.

It is therefore important to study the phenomenon of disaffection in its own right. In order to better understand the phenomenon of disaffection, it is necessary to understand better the notion of affect. That is the quest that drives this study.

Research into Disaffection with School Mathematics

Considering the importance of the issue of disaffection as outlined here, there is not the volume of research that would seem appropriate to the social, economic and individual impact that has been reported in relation to the problem of disaffection with school mathematics. The Royal Society State of the Nation report (2008) into mathematics points out that there has not been enough quality of research into this area, and cites only three studies (Brown et al., 2007; Mathews & Pepper, 2005; Nardi & Steward, 2003) in relation to mathematics. The study by Kyriacou and Goulding (2007) is also relevant, since it is focussed on issues of motivation.

Indeed, a whole-document search on ERIC combining 'disaffection' and 'mathematics' yields just 5 results. Of course, disaffection is referenced in multiple ways and in many more studies than the three cited in the Royal Society report. Almost any study treating issues of attitude or affect will encounter evidence of disaffection. Buxton's (1981) study on anxiety is an important landmark here.[1] In addition, anxiety (particularly maths anxiety) has been much studied over many years (see for instance, Hembree (1990)). However, the three studies cited take disaffection with school mathematics as their primary focus.

Thus affective issues are seen to be important in learning mathematics. Williams and Ivey (2001), suggest that once a student adopts a certain stance towards a subject this then becomes the basis for future action, which in turn can then reinforce the stance taken, forming either a positive or negative loop. Thus, a student that decides mathematics does not interest them may disengage from the subject and make less effort, which will lead to lower achievement and satisfaction. Students can then attribute apparently permanent characteristics either to themselves ('I am not interested in maths') or to the subject ('maths is boring'). Girls are reported to be especially likely to form a fatalistic view of their lack of ability in mathematics as innate (Dweck, 2000).

Negative attitudes to mathematics have been described in detail at Key Stage 3 (age 11–14) by Nardi and Steward (2003) who found that mathematics was perceived as *tedious*, with too much *individual* work and *rote learning*, *elitist,* and *de-personalised (T.I.R.E.D.)*. This was attributed to the concentration on 'teaching to the test' and not enough emphasis on engaging and inspiring students. They identify a "mystification through reduction" (p. 357) effect in which teachers, in an attempt to make mathematics simpler, reduce mathematics to a list of rules and thereby fail to enhance a proper understanding of the underlying concepts.

Nardi and Steward also report a lack of the affective dimension as a perceived characteristic (e.g., in mathematics there are "no positive or negative emotions, you just have to do it. It's like a null period" (p. 361). With an age 16 group, Mathews and Pepper (2005) note that the perception that mathematics is 'dull' is expressed by high-attaining as well as low attaining students.

Both the Mathews and Pepper (2005) and Brown, Brown, and Bibby (2007) studies focus on participation and progression in school mathematics in a post-compulsory context. The Mathews and Pepper study is highly quantitative and focusses on issues of perceived difficulty, and it says little about issues of attitude, belief, motivation or disaffection as such. The Brown, Brown and Bibby study, also is also driven by concern for participation and progression, but it does note the perceived difficulty of school mathematics, and mentions issues of lack of confidence, perceived lack of relevance or utility, lack of enjoyment, and the perceived boring nature of mathematics.

It can be concluded from these studies that mathematics is often perceived as boring and difficult, that teaching and pedagogy help to embed these attitudes and beliefs in pupils, and that these factors combine to create a lack of confidence and a lack of enjoyment in school mathematics.

Much of the evidence cited here is concerned with the incidence of disaffection. The study by Nardi and Steward is the exception in that it goes further than other studies in addressing disaffection directly as an issue of significance for the learning of mathematics, and in trying to characterise the construct of disaffection in research terms. This is particularly relevant to the current study in regard to providing the basis for defining disaffection in such a way that is relevant to the theoretical position adopted in this study, and its methodological design.

CHARACTERISING DISAFFECTION

Disaffection is not an easy construct to characterise in research terms. Part of the reason for this is that (like so many constructs in mathematics education) it is borrowed from a different context.[2] A further complexity arises since it is in turn based on the notion of affect, which in itself is complex, multivariate and poorly understood and structured in research (Hannula et al., 2009).

Dictionary definitions define disaffection as:

> a state or feeling of being dissatisfied with people in authority and no longer willing to support them. (Merrion-Webster)
>
> …describes young people who are no longer satisfied with societies' values. (Cambridge-online)

Other definitions use a range of disparate terms including alienation, disloyalty, unrest, discontent. They point to a social and political bias in the genesis of the term. This has perhaps influenced the way that characterisations of disaffection in educational research have focused on truanting and bad behaviour. Solomon and Rogers (2001) define disaffection (after Kinder et al., 1999) as a sense of being "disengaged and dislocated from…schooling opportunities" (p. 331). They also point out that it is used as an umbrella term for a variety of behaviours, attitudes and values. The causes of disaffection are seen primarily as the school curriculum, being boring, stressful and involving difficult new methods of assessment and grammar school curriculum. Like Kinder et al. they identify relationships between students and teachers alongside these other issues as being important in determining the climate for disaffection. They also point out that disaffection relates to low scores in self-efficacy, and note a lack of sense of agency in disaffected pupils. Nonetheless, it can be seen that behaviour (usually characterised as unwanted or dysfunctional) is often central to views on the notion of disaffection.

Skinner (2008) takes a more operational approach to disaffection. Her model incorporates the influence of context, together with 'self', as defined by the three perceived motivational needs of competence, autonomy and relatedness. These combine to result in internal dynamics that comprises emotion and behaviour, but in an engaged and disaffected form. She claims that this model satisfies much of the known results/data. That is, supportive context, together with positive self-perception fuel engagement in the classroom, whereas less supportive context, together with poor or negative self-perceptions fuel disengagement. This is an interesting model, not least since it seeks to characterise disaffection in terms of motivational variables, but also since it represents a self-system model as a dynamic entity. An important result from this study is the demonstrated relationship between emotional factors such as autonomy and teacher support and behavioural engagement. She reports that emotional disaffection (especially boredom) significantly effects effort, persistence and engagement. One disadvantage of Skinner's approach though, is that it does not take account of cognition, and hence cognitively-mediated constructs such as attitude. It should be noted that the emotional schema is also somewhat idiosyncratic, involving as it does, emotions such as vitality and zest.

Nardi and Steward (2003) offer a number of definitions or at least characterisations from the literature. Their main point is to widen the definition from truancy and disruptive behaviour to include 'quiet disaffection', which is seen as low engagement and perceived lack of relevance. Nardi and Steward also point out that having negative attitudes does not imply an unwillingness to engage with school mathematics. However, the nature of the engagement becomes more like 'resigned acceptance', together with low effort and low aspiration.

Nardi and Steward offer the 'TIRED' framework as a profile of negative attitudes that characterise disaffection. What is interesting on closer examination of these five categories is that they each appear to be a collection of attitudes to, and beliefs about, mathematics. But in their description of these categories there is a conflation of affective factors associated with each. Take for instance 'Tedium'. For disaffected students, school mathematics is seen as boring, and (in Nardi & Steward's description) as lacking relevance, excitement, variety and challenge. This description suggests that disaffection comes about not just because of what mathematics is (or is perceived to be), but also what it is not. This suggests that it fails to meet needs that the pupils have (in this case relevance, excitement, etc.). This in turn suggests that disaffection has a motivational genesis, or at least that motivation is an important component of the condition that needs to be acknowledged. Again, in relation to 'Rote learning', Nardi and Steward discuss the importance of the unmet need for understanding in disaffected pupils. This, again, suggests the importance of motivation, as does the need for a positive view of one's own ability, which is seen as lacking in the 'Elitist' category of the profile. Indeed, the authors, in their review of literature, cite research on motivation as relevant, and describe the distinction between identified and introjected aspects of motivation, suggesting that mathematics is not seen to be relevant (identified), and many pupils engage with mathematics reluctantly, but from a sense of duty to others (introjected).

In a similar way, Nardi and Steward point out that the subjective experience of many of the (perceived) aspects of mathematics, as indicated by the TIRED framework, involve emotional responses for students, such as boredom, anxiety and so on.

The notion of disaffection, then, is fraught with difficulties, complexities and even paradox. Yet some important conclusions can be drawn from the evidence above and particularly that of Nardi and Steward. First of all, it can be seen that the experience of negative affect or emotion does not necessarily in itself imply disaffection. Everyone (or almost everyone) will at some times experience frustration, anxiety or boredom. This is normal and can be unthreatening and insignificant. It is useful here to attend to De Bellis and Goldin's (2006) distinction between local affect (in and of the moment) and global affect (which pervades over time). With global affect, it is possible that the emotion is overwhelming (say anxiety) and ever-present in the experience of learning mathematics. In this case then it would be disabling. It can 'spill' over into all other aspects of subjective experience, including identity. Disaffection is located in global affect in that it endures over time, even though it can manifest consistently or frequently in local affect.

On the other hand, the existence of more stable negative attitudes alone does not necessarily mean that a person is disaffected, as discussed above, although it indicates that they probably are.

According to Nardi and Steward, pupils can engage with mathematics and still be quietly disaffected because the motivation is introjected or identified. Students do mathematics, but they do not enjoy it. In these circumstances, presumably, pupils engage because it satisfies some need (although at the same time clearly not satisfying other needs). An interesting consequence of this is that people may have multiple needs that may be in conflict at any moment in time.

In this way, it can be seen that disaffection is not a one size fits all construct. If a young person is disaffected with school mathematics, the evidence discussed shows that it manifests across a range of affective constructs: motivation, emotion, beliefs (about self and about mathematics), and attitudes. It is embedded in every aspect of their experience and relationship to mathematics, and is not located in any single one of them. What the evidence does show is that to characterise someone as simply dissatisfied or disengaged, is to give an impoverished description that doesn't do justice to the reality of their subjective experience of mathematics.

There does appear to be some overarching aspect of disaffection as manifest in global affect that can be labelled as a lack of attraction. The key to this lack of attraction is that it is characterised by a strong and pervading negativity that can manifest across the range of affective constructs (such as motivation, emotion, attitudes and beliefs). This pervading negativity leads to a reluctance to engage, which in turn disables learning. The evidence also shows that such reluctant engagement (or engagement predicated purely on obligation or duty) also leads to an impoverished experience of learning mathematics. Such delimited engagement seems to bring little sense or anticipation of competence, enjoyment or satisfaction. In this way it can be seen that although engagement is important, the *quality* of engagement needs also to be taken into account.

I take the position that this qualitative aspect of young peoples' engagement with mathematics is not only relevant to the experience of learning, but has not been studied in sufficient depth.

If a person feels that they can't approach or undertake a task because it is too painful, or because they feel that can't do it; or if they ignore the task because it is deemed to be worthless, then learning is prevented or impaired, and this in turn will delimit the development of capability. If a person is thus unable to develop mathematically, then they are disaffected.

The position taken here, then, is that disaffection is characterised by prevailing negative affect that disables or inhibits learning. Not only does the disaffection pervade over time, but it will impede engagement, willingness, persistence and some sense of promised mastery. If this in turn prevents any sense of enjoyment and motivational satisfaction, then this is also at the heart of the notion of disaffection.

Following on from this, there are some things that can usefully be said about disaffection:

- It is not just state or a negative, in-the-moment phenomenon (local affect);
- It pervades over time, but is not necessarily permanent;
- It is characterised by a lack of attraction;
- It involves a pervading negativity;
- It is interwoven with social and cultural issues;
- It manifests across the range of affective constructs;
- It is intimately connected to motivation;
- It involves behaviour like avoidance or disengagement;
- It limits or disables the ability to learn mathematics;
- It precludes a sense of competence, self-efficacy, enjoyment and motivational satisfaction;
- There can be degrees of consequent disablement.

Table 1. Characteristics of disaffection in relation to key affective constructs

Psychological system	*Affective construct*	*Indications of disaffection*
emotion	emotion	Frequent negative emotion Intense negative emotion Sustained negative emotion (mood) 'it gives no joy'
cognition	attitude	Expresses negative attitude Doesn't like or enjoy mathematics
	belief	Limiting beliefs about self Limiting beliefs about the nature of mathematics ('It's too hard', 'I can't do algebra')
	goals	No higher sense of purpose No/low sense of importance/utility for a topic ('I hate fractions') No/low commitment to task in hand ('what's the point of this?'
	attributions	Limiting attributions ('they get more attention.') Low or negative expectations
	self-regulation	Lack of persistence or discretionary effort, inability to deal with setbacks
motivation	interest	Lack of interest or attraction. Boredom Doesn't satisfy needs
behaviour	engagement	Won't engage or avoids mathematics Engagement is reluctant or limited by duty or obligation

It therefore seems likely that the more aspects of affect that are involved, the greater the intensity of the negative affect, and the degree of consequent disablement of learning, will together characterise how disaffected someone is.

Table 1 above, is a compilation of the ways that aspects of disaffection are reported in research to manifest across psychological systems, and across affective constructs within those systems. This tabulation is entirely consistent with research reported above, and includes attitudes and beliefs about mathematics as defined in the TIRED framework, but is not limited by them.

So, whilst the motivational and emotional schema of Skinner et al. and the more attitudinal approach of Nardi and Steward have both made significant contributions in this area, a synthesis that acknowledges the strengths of both will be of value.

Based on this analysis, some conclusions can be drawn that have important implications for this study, and the position adopted. These are:

- Central to consideration of disaffection are the notions of motivation and emotion.
- That young peoples' relationship to mathematics involves multiple affective as well as cognitive variables in a dynamic system, and thus behaviour (e.g., engagement/disengagement) can be seen as both a consequence of disaffection and a dynamic component of it.

Although this makes disaffection a complex and difficult construct, it is no less important, and is deeply rooted in common sense reality. An exploration of the experience of disaffection and the causes of disaffection, and, even more, theorising in relation to disaffection, require us to delve further into the field of affect.

NOTES

1. Part of its importance is in the acknowledgement of the disability caused to adults by fear of mathematics, but also in considering affective issues as central in the learning of mathematics.
2. Usually, social and political contexts, as described below.

CHAPTER 2

THE DOMAIN OF AFFECT

Skemp (1977) showed that he was all too well aware of the importance of affect in learning mathematics. He talks about the importance of maintaining interest and motivation – issues which still concern us today. He understood the notion of 'revulsion' (p. 117) and laid the blame at the door of teachers, their pedagogy and relationship to pupils. He also pointed out the role of understanding in pupil's affective responses to school mathematics. He was perhaps the first to propose that anxiety reduces the efficiency of mathematical thinking. Indeed, Skemp proposes a number of ideas for which serious consideration came only much later. That is not to say that the current study will always agree with Skemp's formulation. To take an example, Skemp seems to associate anxiety with a rote-learning style of pedagogy, and in this he may well be right. He was also clear about the importance of motivation, although his formulation of the construct was somewhat idiosyncratic. He talks about 'optimum level of motivation' (p. 131), but this is to confuse anxiety and motivation. He talks explicitly about intrinsic motivational needs, the need for growth and the pleasure in doing an activity for its own sake.

In his later book Skemp (1979) says:

> In everyday human activity and interaction, feeling and cognition are combined in varying degrees; and this close association between cognitive and affective experiences will follow as a necessary consequence of the theoretical approach which is being developed in this book. The dissociation of the two is I believe an artificial one, which has led to one-sided approaches in both psychological and educational theory. (p. 11)

In a seminal statement of the state of research in affect in mathematics education McLeod (1992) points out that the influence of behaviourism on educational psychology is a major factor in the 'neglect' of the affective domain. He notes that the rise of cognitive psychology has provided a new paradigm which in turn has provided new theoretical and methodological approaches. He proposes the theory of emotions as set out by Mandler as a coherent basis for the study of affect. However, he proposes a structure for affect comprising beliefs, attitudes and emotions, and he makes a strong case that affect and cognition are intimately linked.

McLeod reviews research on a number of other affective constructs, such as confidence, anxiety and self-efficacy. He also considers motivation, noting that "many of the studies that have been discussed in this paper have something to do with motivation" (p. 586). He notes the 'disconnected components' (p. 586), which are seen to make it difficult to integrate motivation as a construct into research. McLeod espouses the need to integrate a variety of constructs into the study of affect, including those that relate affect to the cognitive domain such as metacognition.

In a review on mathematics teachers belief and affect, Phillip (2007), reports that anxiety is studied more than attitude since it is easier to operationalize, and, '(is)... more verbally expressible for students.' This suggests two things. Firstly, that there is an assumption that students are not easily able to articulate their own emotional and affective landscape, and secondly that since emotion is more difficult to research, it has been somewhat ignored. I will take the contrary view that students can talk coherently and articulately about their own emotional landscape, and, indeed, that the voice of students is vital in understanding the issue of disaffection.

Phillip (2007) notes that researchers concluded that students' emotions are resistant to change. This is a widely held assumption in mathematics education research, and will be challenged, in the course of this study and in the evidence that follows. Phillip's review covers the research on anxiety and particularly maths anxiety. One of the key findings is that exams and testing are particularly damaging, and that one of the reasons is that exams focus on time (speed) and recall. Hannula (2002) showed, in the case of Rita, that emotions can change, and that despite her negative attitudes to mathematics, she still put effort into learning mathematics. This begs the question of what else is motivating Rita?

What can be inferred from McLeod's work onwards, and reflected in comprehensive reviews,[1] is an increasing acknowledgement and focus on affect as significant in the study of mathematics education. This is expressed by, among others, Goldin (2003), who says: "when individuals are doing mathematics, the affective system is not merely auxillary to cognition – it is central" (p. 60).

THE INFLUENCE OF ESM 63

It is useful to evaluate how the conception of the field of affect has developed since the time of McLeod by examining more current work. In this sense the special issue of Educational Studies in Mathematics (ESM, 63, 2006), which focussed on affect, is very instructive. In the introduction, (Zan et al., 2006) the authors point out that maths anxiety and attitude to maths are seen as major foci in research. They also note:

> Emotion has been used less in mathematics education research – so far – despite being arguably the most fundamental concept. (p. 116)

And further:

> Motivation has received the most attention among educational psychologists, but remained peripheral within mathematics education. (p. 117)

They note that the six papers all study affect from a variety of theoretical perspectives. These include: affect as representational system (DeBellis & Goldin, 2006); self evaluation and self-regulation processes (Malmivuori, 2006); goals reflected in emotion (Hannula, 2006); somatic markers (Brown & Reid, 2006); socio-constructivist perspective (Op't Eynde et al., 2006); emotions as socially organised (Evans et al., 2006). The issue makes clear that affect is capable of being studied from a variety of perspectives, each of which enables different insights to be brought to the field. For instance, the study of self-regulation not only enables correspondences and relationships between cognition and affect to be studied in a systematic way, but also establishes the influence of cognitive appraisal and consequent adjustment of behaviour as an important component of developing effective strategies for learning and attainment. Socio-constructivist perspectives (Op't Eynde et al., Evans et al.,) also help to account for the influence of context on the learning of mathematics, and particularly on the ways that learning mathematics can be socially determined. Crucially, in the paper by Evans et al. (2006), emotions are seen as having inter-personal rather than individual origins. It will be argued here that by contrast, not all emotions can be viewed in this way, and thus that more individual psychological explanations are needed to give a full account of emotions. For instance, Evans et al. argue that anxiety is caused by reference to positioning in the group, yet, whilst this may be the case at times, it cannot provide an explanation for all occurrences of the emotion of anxiety or panic.

DeBellis and Goldin see affect as a representational system which in turn interacts with cognition. Along with Hannula, they seek to infer affect from behaviour, and it has already been pointed out that this brings dangers of misinterpretation. They posit the notion of 'semi-stable' sequences or pathways of feeling that interact with cognitive configurations. They offer a number of interesting conjectures, which are deserving of further investigation. These include: affective competencies; the notion of meta-affect, which may enable people to experience, for instance, fear as pleasurable.

Malmivuori claims that key concepts in self-systems and self-system processes represent important organising and regulating functions which can explain important results, for instance the relationships between cognition, affect and behaviour. She develops an interesting distinction between automatic regulation and active regulation. Automatic regulation is seen to dominate the evaluation system, and to operate at unconscious or preconcious level. It is seen to hinder higher level mental processing – due to strong negative affect.

These theoretical approaches provide perspectives or lenses through which phenomena can be studied, and it is interesting to see how these different approaches offer contrasting insights and interpretative frames for aspects of affect in learning mathematics.

ATTITUDE

Attitude is perhaps the most widely studied construct in mathematics education. Studies based on attitude have made significant contributions to our understanding of key aspects of learning mathematics. These include an understanding of negative affect, but also gender and other social differences in attitude and achievement in mathematics. On the other hand, concern has also been expressed about research based on the statistical study of attitude. The position in the current study will be elaborated below, but will seek to identify key affordances of the construct, which can then be incorporated into preliminary aspects of the study, but to justify a move beyond attitude for the main study.

McLeod (1992) included attitude in his influential characterisation of the field of affect. Whilst he cites and acknowledges the importance of affective factors in determining outcomes (attainment and achievement), he notes that they react with each other in unpredictable ways. He also notes that emotions have not been a major factor in research in mathematics education, and he attributes this to the difficulty of studying it. The complexity of affective constructs is seen as a barrier to research. So, for instance, he notes that the statistical analyses of questionnaire data "were not necessarily reflecting accurately what students were thinking and feeling." (McLeod, 1994). In this paper, McLeod also suggests that new methods need to be developed that "will help us capture the complexity of the issues" (p. 644). This focus on measurable products, rather than more complex processes, characterises much research in mathematics education. So whilst attitude is an ill-defined construct, it provides easily measured objects, and this in part explains its ubiquity in the research literature.

Schorr and Goldin (2008) argue for "the need to study affect more deeply than the study of attitude permits" (p. 132). And further, "it is increasingly clear that the functioning of affect is far more complex than is suggested by considerations of positive versus negative emotions and attitudes" (p. 133).

One of the problems for studies based on attitude is the lack of clarity and agreement on the construct itself. Ruffell, Mason, and Allen (1998) see attitude as a mental orientation which expresses emotion, but which also consists of a cognitive aspect (beliefs), feelings (emotions/affect) and behavioural intentions. "(We) regard beliefs as part of the cognitive component of attitude" (p. 3). They take issue with the notion of simplistic like/dislike characterisations of attitude and see it as a "a constellation of impulses vying for cognitive attention and triggering physiological and hence emotional responses" (p. 3).

They critique a number of characterisations of the construct as unhelpful, and make a case that:

> Perhaps, 'attitude' is not such a stable and reliable construct; perhaps it is highly influenced by social and emotional context, and personal construction of these. Certainly, if a construct can change so radically in a short time, it is unlikely to provide a fruitful taxonomy for research or for the practice of teaching. (Ruffell et al., 1998, p. 15)

It should be noted that if that is indeed the case with attitude, it would also be true of emotion. In this study, a contrary position will be adopted – that in spite of its volatility, emotion is extremely important to study.

Ruffell et al. also point out other potential weaknesses of the construct, for instance that attitudes clearly don't generate behaviour in any simple, linear or predictable way. They also suggest, quite radically, that attitudes and beliefs may be constructions attributed by teachers to students to make sense of what they observe. This suggestion is also repeated by Zan and Di Martino (2007). What is noticeable is that the notion of intention seems to be quite strongly attached to characterisations of attitude, and this in turn suggests that it is, ultimately, a motivationally-related phenomenon.

Many of these observations are shared by other researchers in the field. For instance, Zan and Di Martino (2007) in their study, note that research on attitude has been judged to be particularly contradictory and confusing, due to the fact that it has given more emphasis to creating measurement instruments rather than elaborating on a theoretical framework.

For the current study, it is also relevant that one of the starting points for this study is what Zan and Di Martino call an 'alarming phenomenon'; the perceived negative attitude of students of mathematics to the subject. For their study, they adopted an interpretative rather than a normative approach to the construct of attitude, and settled pragmatically on a working definition. It does seem, however, that although they do not explicitly state as such, that attitudes are treated as homeostatic, that is, as trait. Three core themes emerged from the study, and these related to emotion ('I like/don't like maths), competence or efficacy ('I can/can't do maths') and belief ('Mathematics is…'). Strong associations were found between liking and being able to do mathematics.

Hannula (2004), in his doctoral thesis talks of discarding the notion of attitude, or at least disaggregating it into different pieces which exist in different parts of his model of affect in order to avoid the problems associated with the construct.

In the current study the subjective experience of disaffection is the primary focus, and this requires an investigation beyond the confines of positive/negative attitude and beliefs. A case has been developed for the primacy of motivation as a means for doing this. However, as the characterisation of disaffection above shows, all affective constructs, including attitude, will need to be accounted for and considered. The view taken here is that attitude is particularly useful and effective in understanding issues around the incidence of negative affect, particularly within and between large populations, and this affordance of the construct will be used in this study.

The Work of Hannula

In this context, the work of Hannula is of particular importance, and his ideas will therefore be reviewed here. Hannula is a current and influential commentator who in turn has reviewed and positioned himself in relation to other credible and respected commentators in the field. His work is informed by the heritage of McLeod and Goldin, but also of the comprehensive work done in the Nordic countries on affect by Malmivuouri, Pekrun, and others. An examination of Hannula's ideas will provide a comprehensive and current review of the field. Hannula's PhD thesis (Hannula, 2004) is an examination of the affective territory, and he comes to the view that affect (more specifically emotion), cognition and motivation are 'an inseparable trio', and as such, they need to be studied together to take account of their interactions. He also views, after Goldin, emotion as a separate but connected representational system to cognition. This is in contrast to others (Op't Eynde et al., 2006), as well as Vygotsky (1986) who view affect as primarily grounded in and defined by social context, and that social positions are defined within the Zone of Proximal Development. Hannula concludes that emotions are closely related to goals, and thus motivation is invoked as a key construct. In a radical turn, he rejects both attitudes and beliefs as sound affective constructs, but relevant aspects of these constructs are accounted for by consideration of cognition (beliefs) and motivation (values and attitudes), and this study is in agreement with this position.

In terms of motivation, Hannula focuses on states and processes, and rejects the notion of trait. Of trait, he says:

> In most contemporary research, the focus has been on motivational traits; Such research may help us predict future learning orientation and success, but it will not help much in understanding why a particular student is putting a lot of effort into some activities and not into others or how to induce a desired motivational state in others. (Hannula, 2004, p. 23)

Hannula conceptualises motivation as a potential which energises and has direction. He further explains that it is not a potential for behaviour, but to direct behaviour or behavioural choices. He sees it as intimately connected to emotional control mechanisms, and as manifested in cognition, emotion and behaviour. From here he turns to the needs that underlie motivation, and here, he adopts the widely held view (after Deci & Ryan) that the key needs in an educational setting are autonomy, competency and social belonging. Hannula describes that these needs are then processed and mediated by cognition to produce goals.

In terms of emotions, he notes the distinction between different traditions which see emotions as physiological, the Freudian tradition which sees emotion as grounded in instinctual drives, and the cognitive approach, which argues that emotions are a psychological phenomenon, guided by cognitive appraisals. His analysis leads him further to say that emotions 'code information about progress towards goals' (Hannula, 2000, p. 27). He thus attaches emotions to goals, and this may be too specific, although it seems reasonable that emotions relate to needs, and to progress (or lack of it) towards goals.

Hannula also considers the literature on self-regulation. He sees self-regulation as a system concept that refers to the management of our behaviour through various control systems such as attention, metacognition, motivation, emotion, action and volitional control. In the picture of affect that is emerging through the work of Malmivuouri, Boekaerts, Hannula, and others, self-regulation is an important addition to the construct-space around affect. Not least, it provides a mechanism for considering the interaction between cognition and affect (or affective constructs). Although it is a system concept, it has shown that it is capable of being operationalized, and further, of enhancing our understanding of relationships between cognitive and affective constructs and outcome variables such as achievement.

Hannula notes that affect is located in the world of subjective experience, although it cannot be ignored as a social text or as physiological reality, and that the constructivist paradigm is appropriate for research into the world of subjective experience, and this position is strongly reflected in the current study.

Hannula's contribution to the ESM special issue on affect (Hannula, 2006) begins with a bold statement – "To understand student's behaviour we need to know their motives" (p. 165). He says, in critique of mainstream motivation research, that there should be a focus on states and processes as well as traits. They (traits), he says, will not help us to understand why a particular student is putting a lot of effort into some activities and not into others. These comments seem to me to be self-evident, and I would endorse all of those statements. He sets up a simple model; needs lead to goals, and goals lead to means.

Hannula goes on to suggest that although emotions and cognition are only partially observable (and even partially inaccessible to the person himself), behaviour is always "a dependable manifestation of motivation" (p. 167). But surely, this is not universally true. Consider a relatively common behaviour of truancy or school avoidance. What motivation is manifest in this behaviour? For student A, they may avoid school because they are being bullied, and school is an experience fraught with anxiety and apprehension. For student B, they may avoid school because they are bored. On the other hand, for student C, they are not a 'serial avoider', but sometimes truant because of peer pressure to do so. In each case, the motivations are different, even though the behaviour is ostensibly the same. So, Hannula's statement that 'inferences of the unconscious and subconscious can be made from behaviour', is challengeable.

More recently Hannula has attempted to provide coherence to the field of affect (Hannula, 2012). He points out that there is no shared language or theoretical framework to enable systematic study of the affective domain. He states that attitude as an emotional disposition was typically seen through a positive/negative duality, which misses important distinctions. His analysis concludes that we need richer theoretical frameworks that show relationships between constructs, and takes account of both state and trait. He thus proposes a multi-dimensional framework that has cognition, motivation and emotion on one axis, state and trait on another, and physiological, psychological and social on the third axis. He then allocates constructs to these various cells in the structure. This is perhaps one of the first, or at least one of the major attempts to place state in equal importance to trait and homeostatic considerations in the theorising of affect.

Hannula calls this classification system 'meta-theoretic'. And although classification systems have the potential capacity to provide explanatory value (like the periodic table of elements of Mendeleyev), there is not yet a suggestion that this one has such capacity. Although it provides for connection between, for instance, motivation and emotion at a metatheoretic level, it leaves unsaid the actual mechanisms of such connections.

With much that Hannula has written, his approach takes us beyond some of the limitations of the framework proposed by McLeod. The position taken here is in agreement that affect is driven by motivation and emotion, and that it is connected intimately to cognition and to self-regulation, and that they operate co-dependently. Equally, the distinction between state and more stable constructs and phenomena is a useful one, although the systems approach has a better way of describing these than does the notion of trait.

In terms of Hannula's meta-theoretic schema, it can be seen that the current study is located in state more than trait (although considerations of stability over time will also be addressed); it is concerned with psychological more than physiological or social considerations; This study will also privilege motivation and emotion over cognition, although the latter cannot be ignored.

However, I agree with Hannula that cognition, motivation and emotion, together with notions of self-regulation, provide the core of the affective system, although affect as a 'social text' cannot be ignored. Hannula sees motivation as a potential which provides energy and direction for experience, but I will argue that the qualitative nature of motivation is an important consideration which is too little researched, but which has substantial relevance for the study of disaffection. I agree with Hannula that motivation relates to needs, but the current study will not be confined to the three needs of competence, autonomy and relatedness, but rather an alternative formulation will be derived from Reversal Theory. Even so, a further alignment between the current study and the approach of Hannula is that a constructivist paradigm is here seen as appropriate for research into disaffection since it privileges the world of subjective experience.

MOTIVATION

Motivation appears to be somewhat of a 'Cinderella' construct in mathematics education research. In his landmark paper on affect, McLeod (1992) writes just two paragraphs on the subject of motivation. He says "One of the difficulties with the work on motivation is the diffuse and disconnected nature of the field" (p. 586). He quotes Norman (1981) approvingly, suggesting that 'motivational factors' could be explained as being derived from beliefs and emotions. This is somewhat inconsistent with the notion of motivation as a driving force; one that is the source of our intentional action.

Weiner (1990) plots the history of motivational research from the 'mechanists' and machine metaphors of the early 20th century, through to the cognitivists. From those early days, motivation had become associated with drives, and with rewards and reinforcement, which is the classic 'operant conditioning'. The influence of the notion of rewards is present in the work of Deci and Ryan (2001a) and the Self Determination

Theory (SDT) which derives from it. Again, educational phenomena (in this case rewards) are studied experimentally because success and failure could be manipulated in the laboratory. Motivation is associated with achievement striving, which is seen to be the determinant of behaviour. With the rise of the cognitivist approach, attention shifts to attributions, anxiety, self-esteem, control, intrinsic and extrinsic motivation. "There is an increasing range of cognitions documented as having motivational significance" (Weiner, 1990, p. 620). Weiner notes that grand theories have faded away, although achievement striving remains at the centre. Weiner notes:

> There is an abundance of evidence that motivation influences a vast array of other variables...educational psychologists must broaden their nets to capture the richness of motivational impact. (Weiner, 1990, p. 621)

Middleton and Spanias (1999) say:

> Simply stated, motivations are reasons individuals have for behaving in a given manner in a given situation. They exist as part of one's goal structures, one's beliefs about what is important, and they determine whether or not one will engage in a given pursuit. (p. 66)

This is a very cognition-driven definition. It hints that motivations may be post-hoc rationalisations for behaviour. It also suggests that motivations are determined by goal structures, yet it leaves unsaid where the goals come from in the first place.

> Students who are intrinsically motivated engage in academic tasks because they enjoy them ...they feel that learning is important. (Middleton & Spanias, 1999, p. 66)

This definition can be challenged on a number of grounds. First of all, although people may be motivated to perform tasks, and although they may wish to enjoy them, it can't be assumed that they are always motivated to a task because of this enjoyment. A pupil may be motivated to a task to please a parent or a teacher. Equally, students may be motivated to a task because they feel that learning is important, but they may also be motivated for other reasons entirely.

Pintrich (2003) in a paper discussing the position of research on motivation notes how motivation has moved to the centre of research in understanding achievement and engagement, albeit he says that the field is also fragmented and diffuse. He notes, along with others, the move away from explanations focussed on drive to constructs which focus on cognition and regulatory constructs, and in particular social-cognitive models which involve constructs such as perceived competence, attribution, self-regulation, self-esteem. He also notes that a weakness of these models is their reliance on cognitive and rational processes, which see motivation as 'cold' and that affect (and the emphasis on needs) has been somewhat pushed out. Pintrich points out the proliferation of socio-cognitive constructs that relate to motivation: self-efficacy; attributions and control beliefs; interest and intrinsic motivation; value, as in task value; expectancy; utility; importance; goals.

In the latter context he notes:

> Future research on achievement goals needs to move beyond a simplistic mastery goals (good) versus performance goals (bad) characterisation. (Pintrich, 2003, p. 676)

He goes on to say: "Simple one-shot correlational studies with self-report instruments will probably not provide us with much more knowledge gain" (p. 678). To do so, he claims, will involve the use of creative new methods to access the more implicit and unconscious aspects of motivation.

> The role of affective factors, including both general moods and specific emotions, are not well understood and have often been ignored in our current social-cognitive models of motivation. With the exception of anxiety there has been little study of the role of emotions. (Pintrich, 2003, p. 679)

Hidi and Harackiewicz (2000) discuss the role of interests and goals in relation to unmotivated students. They note evidence of a whole range of sources of situational interest including novelty, surprise, vividness, intensity, as well as utility or personal relevance. They further note that working with others increases situational interest. They comment on the dominant theoretical approaches of SDT and of Dweck and the mastery-performance goal distinction. In terms of the latter, they judge that they are not as dichotomous as they appear, and that they co-exist and interact in multiple goal contexts.

In terms of SDT, Hidi and Harackiewicz argue that children exhibit patterns of motivation that mix extrinsic and intrinsic. They note that "emerging perspectives in the literature now suggest that academic motivation should not be evaluated on a simple intrinsic-extrinsic continuum" (p. 164).

Self Determination Theory

Given the strong influence that behaviourism has had over educational psychology in the middle of the twentieth century, it is perhaps not surprising that one of the key notions studied in educational psychology was that of rewards. In a series of experiments in the 1970's, Deci, Ryan, and colleagues found interesting and unexpected results. This was that rewards, when added to a situation that is already rewarding in itself (now labelled intrinsically motivating), reduce motivation, rather than increase it. These results were something of a paradox. These results led to the development of the notion of intrinsic motivation, although the idea goes back to the work of De Charms.

In early experiments by Ryan and Deci (1971), students were invited to take part in experiments, for which some would receive rewards, and some would not. Interestingly, and paradoxically, motivation was shown to be stronger for the groups who were not going to be rewarded. That is, taking away the reward increased motivation. So developed the notion of intrinsic motivation. Where motivation is governed by reward or external pressure it is labelled extrinsic.

Niemiec and Ryan (2009) after Ryan and Deci (2000) define them thus:

> Intrinsic motivation refers to behaviours done in the absence of external impetus that are inherently interesting and enjoyable. For example, when people are intrinsically motivated they play, explore, and engage in activities for the inherent fun, challenge and excitement of doing so…accompanied by feelings of curiosity and interest… Extrinsic motivation refers to behaviours performed to obtain some outcome separable from the activity itself. (pp. 134–135)

Extrinsic motivations are seen to become more effective as they become more internalised by the subject. They point out that not all aspects of school learning are inherently interesting or fun, and if intrinsic motivation is not evident, other incentives need to be introduced. However, SDT sees education as providing external impetus to learning in the form of controls, evaluations, rewards and punishments (extrinsic), that can be damaging to young people's natural tendency to learn.

In Self Determination Theory, the basic needs that can be satisfied by intrinsic motivations are seen as being competence, autonomy and relatedness.

> The need for autonomy refers to the experience of behaviour as volitional and reflectively self-endorsed. (Niemiec & Ryan, 2009, p. 8)

And further:

> Autonomy refers to being the perceived origin or source of one's own behaviour. (Ibid., p. 8)

The need for competence refers to the experience of behaviour as effectively enacted. Competence is seen to refer to feeling effective in one's ongoing interactions with the social environment and experiencing opportunities to exercise and express one's capacities.

> Relatedness refers to feeling connected to others, to caring for and being cared for by others, to having a sense of belongingness both with other individuals and with one' community. (Ryan & Deci, 2002, p. 7)

These needs are seen to foster the most volitional and high quality forms of motivation, and the degree to which they are thwarted will prove detrimental to intrinsic motivation.

These constructs, they report, have been supported by numerous empirical studies. Certainly, the notion that autonomy-supportive (rather than controlling) teaching enhances interest is an important result in the field. These basic psychological needs are seen as the key to understanding the unexpected behaviour identified in the studies reported above. External events can enhance or diminish intrinsic motivation according to whether they are perceived as enhancing competence and self-determination or diminishing them (Deci et al., 2001b).

Deci and colleagues found that offering people extrinsic rewards for behaviour that is intrinsically motivated undermined the intrinsic motivation as they grow less interested in it. Initially intrinsically motivated behaviour becomes controlled by external rewards, which undermines their autonomy. Tangible rewards are perceived as controlling and will thus decrease intrinsic motivation. Tangible rewards have negative consequences for subsequent interest, persistence, and preference for challenge.

Despite its importance, it is possible to challenge some of the assumptions of SDT, but this will be done, below, from the perspective of Reversal Theory, once that theory has been introduced.

EMOTIONS

The intimate link between emotion and the experience of learning mathematics has been recognised for a considerable period of time (Buxton, 1981). Buxton's study provided evidence of the damaging personal affect associated with lack of confidence and competence in mathematics. Skemp (1977) also acknowledged the part that motivation and emotion (anxiety and pleasure) play in the learning of mathematics, and attributed anxiety to poor or instrumental teaching methods. He went on to propose the derivation of a range of emotions, linked to successful or unsuccessful strategies for learning mathematics (Skemp, 1979).

Despite this early attention, emotion seems to hold an unusual place in the literature on mathematics education, and even in the literature on affect in mathematics education. Whilst it gains attention as a topic of interest, there is little systematic or detailed data on emotions and alongside this, the construct is not often fully theorised.

In much research where emotion is reported, it is characterised as simple positive or negative emotion. The most frequently mentioned aspect of emotion is anxiety, and more specifically test anxiety. Yet there is also some weight of opinion that there is a need to go beyond simple positive versus negative characterisations (Hannula et al., 2009). There could be a number of reasons for this. It appears to be more difficult to build a sound theoretical foundation for emotions that relate emotional perspectives and phenomena to other affective constructs. Another possible reason is that, in the search for cause and effect relationships, attitude or beliefs may seem to offer more fertile territory for finding such effects since they lend themselves to quantitative analysis via survey-type methods.

So while there appears to be little attempt to build hypotheses based on cause-effect principles involving emotional variables, emotions do have an important part to play in a) understanding the impact of low attainment and disaffection with mathematics and b) understanding the mutual interaction between emotion (as well as other affective variables) and learning mathematics.

In studying affect and emotions, McLeod's (1992) formulation, with its approach to emotion inherited from Mandler, has been highly influential. This characterised emotion as 'hot', unstable and short-lived. Attitudes and beliefs were seen as more stable, and thus closer to trait. This in turn implied that these affective constructs could be measured quantitatively. This approach has been so influential that it has seen emotion also treated as trait, for example Givvin et al. (1996) who measured simple, enduring positive or negative emotion.

Mandler's theory is an arousal-cognition model of emotion. Emotion is seen to arise from the interruption of an individual's plans. There is perceived to be an incongruity between what is expected and actual events. A hard-wired response to interruption is 'activation of physiological systems.' (Allen & Carifio, 1995). This arousal is seen to determine energisation or intensity. Cognitions (in the form of appraisals) and arousal are seen as key aspects of emotional response. Part of the value of Mandler's contribution is to link emotion to learning, mathematics education and problem-solving in particular. He also characterised emotions as agreeable versus disagreeable, thus suggesting that hedonic tone and the bi-polar nature are important aspects of emotions. However, the evidence is equivocal. Different studies find both positive and negative relationships between emotion and arousal (Allen & Carifio, 1995). And this in turn points to a possible major weakness in Mandler's approach in that it fails to account for low-arousal emotions such as boredom, relief or relaxation. Nonetheless, emotion is seen as complex and non-linear, as a system, is seen to interact with cognition and motivation systems. This makes it a challenge to theorise, and indeed, to study.

Some researchers have gone beyond simple positive and negative characterisations (Pekrun et al., 2007). But Pekrun considers emotion as related directly to achievement activities and outcomes, as do Anderman and Wolters (2006) who define it purely in terms of goal orientation. These characterisations do not account fully for the wide range of emotions we see reported in other research (Schorr & Goldin, 2008). Pekrun et al. (2007) identified a range of positive and negative emotions in their qualitative studies, although anxiety was by far the most common. This may, however, be because the emotion is studied in testing situations. They state that "emotions influence students' cognitive processes and performance as well as their psychological and physical health" (p. 92).

One point of interest is that they noted the existence of meta-emotions, such as the ability to be able to cope with anger. Indeed, their theoretical approach, component process modelling, suggests, along with others, that emotions are sets of interrelated psychological processes that include the affective, the cognitive and physiological as well as the motivational, and, as mentioned above, self-regulatory processes such as meta-emotion. Their findings show that emotions relate in significant ways to students' learning and achievement. Further, although they find that emotions can be distractive of efficient task engagement, they can also act in more positive ways. Enjoyment and boredom appeared to be the most influential emotions in both the positive and negative manner.

Another approach that views cognition, motivation and emotion as the key processes involved in learning is that of Schutz and DeCuir (2010). But whereas Pekrun et al. seem to view attribution and expectancy as the key cognitive variables, Schutz and DeCuir see goals as "providing direction for our thoughts, behaviour, strategies" (p. 127). They assert that appraisals result from beliefs and personal theories and relate to our progress towards goals. And although both see emotion as of equal status and importance for learning, both view cognitive appraisal as the driving force for emotion. Thus they can be said to be cognition-led. But does well-formed reasoning about goals really inform our emotion responses as much as is implied in these approaches? Surely, we are hard-wired, when faced with physical threat to respond both physiologically and emotionally. Fear and severe anxiety, for instance, may arise from more unconscious and possible pre-cognitive processing. It may be more accurate to say we are responding to needs, since goals imply a higher degree of conscious and rational processing than may be the case in many situations.

However, Schutz and DeCuir do point out the tendency of research to characterise emotion as trait, and that this in turn tends to promote more reductionist interpretations. On the other hand, their approach also overemphasises the importance of the social-historical and cultural context, and the view that goals are purely a manifestation of context does not explain that different individuals respond differently to the same context, or indeed, that an individual can respond to the same context in different ways at different times.

In their taxonomy, emotions are categorised in terms of ‘activation’ (low or high) and value (positive and negative). However, as with most current theories of emotions, there is no account of how an individual might traverse through these emotions, or how they transform in sequences from one to another.

Although not specific to mathematics, Meyer and Turner’s (2010) review of emotions in classroom motivation research also argues for the key role of emotions in learning. At various times, they use the words ‘essential’, ‘pivotal’ and ‘central’ to describe its emerging importance in their work. One interesting new perspective in their work is the understanding of the importance of affect in teacher discourse. They claim that teacher’s affective support for well-being is a vital determinant in classroom learning. They also point out that emotion is often treated as an outcome variable in motivation research, and that too little is known about how emotion influences behaviour and other responses. They argue that it is necessary to go beyond trait in these considerations.

The need to treat emotion along with cognition and motivation as dynamic systems which interact gains an explicit treatment of emotion in a dynamic component systems approach (Op’t Eynde & Turner, 2006). They also argue that moods and emotions are not side effects but integral with motivation and cognition in learning. They define emotion as “a relatively brief episode of coordinated brain, autonomic, and behaviour changes that facilitate a response to an external or internal event of significance to the learner” (p. 362).

Op’t Eynde and Turner identify similar dimensions of appraisal to other researchers, but pose an interesting idea in the notion of ‘coarse processing.’ This refers to the pre-conscious processing that may happen before more higher-cognitive processes, and that involves noticing or shifts of attention that may have consequences or personal impact. They also note that the research requires a shift of focus from observer to actor perspective, since what matters is appraisals, interpretations and meanings for the student. As with other socio-constructivist approaches, the characterisation of emotion as exclusively social in nature is a limitation of this approach.

There have been attempts to account for disaffection using motivational and emotional constructs, but these have applied in the general case of school education, rather than specifically in mathematics education. One such example is Skinner et al. (2008). They focus on the notion of engagement, which they see as having both behavioural and emotional components. They view disaffection as negative engagement, or, more specifically: "the occurrence of behaviours and emotions that reflect maladaptive emotional states." (Skinner et al., 2008, p. 767). Amongst the emotions accounted for are boredom, anxiety, anger and shame. The framework takes account of these emotions, but doesn't give a theoretical account for their genesis.

Evans (2000) claimed "there is still little or no explicit acknowledgement of the importance of the affective – feelings of anxiety, frustration, pleasure, and/or satisfaction which attend the learning of mathematics" (Evans, 2000, p. 108). However, this is beginning to change. Evans in turn was influenced by Walkerdine (1988), who used ideas from psychoanalysis to inform her understanding of emotion in mathematical discourse. Power is seen as pleasurable and in this Freudian view, repression and fantasy are key concepts. Early (1992) also adopts a psychoanalytic approach in his study of fantasy images prompted for his research. However, amongst the limitations of the Freudian approach are the necessary adoption of constructs such as repression and fantasy, which have yet to prove their utility in this context. Another limitation is the narrow range of emotions addressed, being focussed primarily on fear and anxiety. Breen (2000) also focusses on fear and anxiety and describes vividly the frightening picture of the damage caused by students' mathematical experiences (even of those who had obtained good passes in mathematics at the school leaving age). He laments the scarcity of research in this area.

Another, more recent trend in research related to emotion and learning is the neurophysiological one. An example is Caine and Caine (2006) who focus on executive functions of the brain, and in particular, how certain circumstances can sabotage optimal functioning. Here again, much of the focus is on fear and panic and maladaptive responses to high arousal.

Schutz and DeCuir (2002) point out that there are methodological problems in the attempt to study emotions. They note that over the millennia, and up until the present, emotions are seen as in conflict with reason (cognition). Even now, in much research, emotions (particularly arousal) is seen as interfering with proficient cognition. They see action as intentional or goal-directed, and they say, goals are manifestations of social-historical context. "Goals provide direction for our thoughts, behaviour, strategies" (p. 127).

At the same time, they do see considerations of trait as an inhibition as it "tends to promote reductionist investigations of intention" (p. 129). They conclude that there needs to be a more holistic approach to researching the cognition-motivation-emotion nexus. It follows from this that one cannot properly study motivation without studying emotion. They introduce the notion of emotional intersubjectivity, which is an interesting notion that suggests that emotion has to be considered in the study of the classroom context, and that this might shed light on issues of gender or social class differences in achievement.

Like Schutz and DeCuir, Meyer, and Turner (2010) see emotion as both important and under-researched in terms of teacher-pupil discourse. Indeed, they cite a body of research evidence that supports the notion that teacher affective support for well-being is a vital determinant in classroom learning. In particular, they see issues of risk, enjoyment and flow as not sufficiently considered. Meyer and Turner also make an extremely important point, and one that flows from a holistic view of the cognition-motivation-emotion nexus, and in particular when this is viewed from a systems perspective, and that is that emotion is traditionally seen as an outcome variable. But in a systems view, it also has to be accounted for as an input variable, i.e., it can in some situations be seen to be a driving variable in understanding transactions.

This view of the cognition-motivation-emotion nexus from a systems perspective is given a systematic treatment in Op't Eynde and Turner (2006). Thus they see moods and emotions not as side effects, but as integral to the study and operation of motivation and cognition. Like other commentators, they try to put some structure to the notion of appraisals and to suggest criteria that humans use in appraising situations. These include novelty, pleasantness, significance, and control. These in turn might give us some insight into the motivational needs and drivers. That is, it is not always situation that is important, but our appraisal (meaning) of the situation, and this in turn may be influenced or even directed by our motivational state and the particular motivational value operational at the time.

The review above leads to three broad conclusions. The first is that emotion is an inseparable aspect of disaffection, and to understand more about disaffection it is necessary to understand its emotional correlates. This leads to the second conclusion, which is that in order to do this, a phenomenological frame, which focusses on the subjective experience and meaning-making of students is perhaps the most important perspective. Although negative emotion has, in the context of quantitative studies been treated as homeostatic (trait), it also seems clear that research needs to acknowledge that minute-to-minute shifts in emotion need to be examined. Thirdly, it seems clear that it is important to widen the investigation to give equal attention to a whole range of emotions in addition to fear and anxiety, that are relevant in the lived experience of disaffected students.

It will be argued, below, that Reversal Theory offers a theory-led and comprehensive account of emotions, that can provide a coherent framework for studying emotional phenomena in a mathematics education context.

NOTE

[1] Leder and Forgasz (2006) give a good historical account of research in affect both within Conferences of the International Group for the Psychology of Mathematics Education (PME) and beyond the PME community.

CHAPTER 3

REVERSAL THEORY

A number of conclusions can be drawn from the analysis set out so far. These conclusions can be woven together to make a coherent argument, and to arrive at a position that in turn frames this study. To summarise some of the key points:

- Disaffection with school mathematics is a serious problem with individual, social and economic consequences.
- It has been too little studied and researched.
- Inasmuch as it has been researched, that research has been dominated by the quantitative study of attitude. Not enough is known about the subjective experience of disaffection. However, moving away from the quantitative study of attitude requires some creativity, and the development of innovative methods and instruments to elicit high quality data.
- Disaffection is a complex and multi-dimensional phenomenon. However, it is, by nature, located in affect, although it has correlates in all affective and cognitively-mediated affective constructs such as motivation, emotion, attitudes, beliefs, interest, engagement, and including self-regulatory and metacognitive skills.
- It has been argued that motivation and emotion form the core underpinning of affective experience, in the sense that they are prior to other affective variables, and influence how we experience the world, but also how we develop other affective dispositions (such as attitudes), and how we interpret our experiences through them.
- Emotion, particularly, is increasingly seen as more central to our experience of learning mathematics, yet it remains very little studied. Studies have focussed on anxiety, or simple positive/negative distinctions, and the full range of our emotional landscapes have not been explored in detail. There is a lack of robust theoretical frameworks that account for the full range of emotions experienced in mathematics classrooms.
- Mathematics education research should be open to new theoretical frameworks that can offer new approaches and insights into these difficult and complex problems.

I have argued that there is a need for new approaches to motivation and emotion that take account of the dynamic nature of experience, which can provide sound theoretical underpinnings to the integrated study of personality, motivation and emotion. By examining new frameworks, not only can new light be shed on current results that are perplexing, but new insights can be generated. In addition, new theoretical perspectives will in turn generate new methods and approaches to research that can push the boundaries compared to the current repertoire of approaches. Reversal Theory (Apter, 2001) is one such candidate for a viable and robust framework.

REVERSAL THEORY

The theory is a comprehensive account of personality, motivation and emotion, and as such, affords a basis for understanding affect centred on an individual's subjective experience. The theory was developed 30 years ago as a way of trying to explain and illuminate the problematic behaviour of young people. One of its primary premises was that the same behaviour could have very different meanings for different people. In an educational context, it was found that the behaviour of school avoidance could have conflicting motivational and emotional rationales. One child might avoid school because they are feeling bullied, intimidated and fearful, whilst another child might avoid school because they are bored and lacking stimulation.

This insight led to a focus on the subjective experience and the meanings ascribed by children to their own behaviour. It was further found that not only was the same (or similar) behaviour given different meanings by different people, but that alternative meanings could be relevant for the same people at different times. This led to the idea that some aspect of their subjective experience was dynamic and changing over time. However, it was also found that subjective experience is not random but has structure.

The literature provides substantial evidence to support the theory and its associated constructs in many and varied areas of human experience. Reversal Theory has proved itself extremely useful in providing insights and explanations particularly in problematic and contradictory areas of human behaviour such as psychological and sexual dysfunction, sport (Kerr, 1999), addiction, risk taking, soccer hooliganism, truancy and juvenile offending, and many more. Much of this evidence is summarised in Apter (2001).

There is also some evidence relating Reversal Theory to the educational context. Some of this is to be found in the proceedings of the bi-annual Reversal Theory Conference (founded in 1975). A pragmatic description of Reversal Theory ideas in an educational context can be found in Mallows (2007). Other evidence relating the theory to education comes from the Learning and Skills Council in the UK. In their report into Learning Styles (Coffield et al., 2004), the potential power of the theory (and associated psychometrics) to understand how students engage with learning was recognised.

Research by Svebak (1993) has shown that an emotion-based psychometric instrument can predict academic outcomes, and is also relevant in this context because it relates affective variables (in this case, emotions) to educational outcomes.

Collectively, this prima facie evidence provides strong support and a coherent backdrop to the case for investigating the implications of the theory in the educational research context.

Overview of the Theory

An overview of the basic principles and concepts of the theory as it has now developed will be given, since this will be relevant to understanding the theoretical position and interpretative framework of this study. Apter (2001) will be used as the key source.

Motivational style is defined as a "distinctive orientation to the world based on a fundamental psychological value" (p. xi). Methodologically, Reversal Theory is structural phenomenological, in that it begins with subjective experience, but the range of that experience is seen as highly structured and ordered into four fundamental domains. The first is called means-ends, and is about directionality or purpose. Put in everyday terms, it is about the journey and/or the destination. So much of psychological theory on motivation is about goal-seeking action and behaviour, and that is encompassed in this domain.

The second domain of experience is about rules. That is, rules as required not just by the law, but by expectations, conventions, norms, customs, and the constraints put on us by social contexts of various sorts. The third domain is one of transactions, which represents those people or things we interact with. Finally, the fourth domain of relationships define us by separation or identification with or from external entities – be they people, teams, or other more abstract entities such as nations (or even 'the world', or 'nature').

One of the key insights of Reversal Theory is that we can experience each of these domains in two, entirely opposite ways. A basic proposition is that high arousal can be experienced in two different and mutually exclusive ways. Take the example of a fairground roller-coaster. We can either be aware of the extreme danger and be focused on outcome or consequences (called the serious or telic state), in which case we are likely to feel anxiety or even panic. Or we can revel in the experience and enjoy the 'buzz' of the momentary thrill (the playful or paratelic state), in which case we will experience the situation as excitement. What's more, we can reverse between these two different states. In a mathematics education context, we might experience both of these states (but not at the same time) when faced with a problem or puzzle to solve. We might be overwhelmed by anxiety or panic at the thought of failure (serious), or immerse ourselves pleasurably into the activity itself (playful). This latter way of experiencing bears a great deal of relationship to the state of 'flow' as described by Czsikszenktmihalyi (1996). What Reversal Theory states is that one will always (at any one moment) experience the world as either serious (telic) or as playful (paratelic), but never both at the same time. It also says that we will move (or reverse) between the two states at frequent intervals. The two ways of experiencing each domain are shown below.

Table 2. Motivational states

Domain	*State*	*State*
Means-ends	Telic (serious)	Paratelic (playful)
Rules	Conformist	Negativistic (rebellious)
Transactions	Mastery	Sympathy
Relationships	Autic (self)	Alloic (other)

When a person is in a state, it colours every aspect of their felt experience, from what they are paying attention to, to what they value, how they view events, and the emotions they experience. More specifically, each motivational state has associated a value, a feeling, a way of experiencing, and associated emotions. These (apart from the emotions – see ahead) are summarised in the table below.

Table 3. Motivational states and feelings

State	*Core value*	*Desired feeling*	*Way of experiencing*
Telic	Achievement	Hi significance	serious
Paratelic	Fun/enjoyment	Lo significance	playful
Conformist	Fitting in	Low negativism	conforming
Negativistic	freedom	Hi negativism	challenging
Mastery	Power/control	Hi toughness	competitive
Sympathy	love	Lo toughness	affectionate
Autic	individuation	Lo identification	Self-oriented
Alloic	transcendence	Hi identification	Other-oriented

As well as influencing what we see and how we interpret the world, they also influence our actions and behaviour.

Within Reversal Theory, motivation is seen as fundamental and ever-present in human experience.

> Motivation enters into and provides a continuing internal context for all of our perceptions, thoughts, and actions. It underlies the very structure of experience. (Apter, 2001, p. 5)

Technically, the motivation in Reversal Theory is meta-motivation. That is, it is *about* motivation or different ways of interpreting aspects of our experience to do with motivation. Thus it is one level abstracted from the content of reality. For instance, although we may know that a person is in the serious state, this does not uniquely define or tell us what the goal or outcome is that they are focusing on.

It should be pointed out that the term state, as used here, is not used to depict a particular emotion or mood, but rather a fundamental psychological motive that orders the whole structure of experience at that point, and that affords a range of possible specific emotions.

The basic principles of Reversal Theory, derived from substantial empirical research over the years have established that:

- We are in one of each pair of states at any one time (i.e. 4 states).
- That one (or possibly two) of these four states will be focal at any one moment. That is, it (they) will operate at the forefront of our phenomenal field, and thus will be more influential in driving our experience of the world at that moment. This focus can change.
- That we will reverse between states frequently. Indeed, it is known to be psychologically dysfunctional to be stuck in a state, and unable to reverse.

It can be inferred from this that people move between states (and combinations of states) frequently. And since these states are opposites, not only are people bi-stable, but they are multi-stable. That is, one can literally, be different people at different times. In this way, Reversal Theory captures the dynamic, changeable (and even contradictory) aspects of our personality, emotions and motivation.

These concepts are antithetical to trait, but the semi-stable aspects of our distinctive personalities are captured in the theory.

Critique of Reversal Theory

Reversal Theory has been somewhat ignored by the wider field of psychology. Perhaps one reason for the lack of a wider profile is that it challenges a number of long-held assumptions such as trait, and introduces new constructs such as the notion of reversal itself. In contradistinction to the notion of trait, Reversal Theory states that we will be different people at different times, and that it is psychologically healthy to be so.

This is not to say that Reversal Theory is free from criticism or challenge. Cramer (2011) has offered a comparative evaluation of the theory based on six criteria:

- Comprehensiveness
- Precision/testability
- Parsimony
- Empirical validity
- Heuristic value
- Applied value

Cramer reports that none of the major theories of personality that he evaluates are without weaknesses, in some cases despite their wide use and influence. He rates the theory as high or very high on comprehensiveness, parsimony and applied value (due to a range of published work on business, sport, health and wellbeing, psychological dysfunction, and so on). By contrast, he rates the theory as poor on precision/testability and moderate on empirical validity. Precision/testability is seen as poor because of the difficulty of measuring states in flux and because there are too few predictive studies which test clear hypothesis based on the theory.[1] As examples of this he cites work needed on the measurement of reversals and on the internal consistency of measures in order to address current weaknesses, and to add credibility to the theory within the wider scientific community.

Apter (2013) gives an evaluation of the current state of the theory from the perspective of further research that needs to be undertaken. He notes that much research undertaken using the theory has been 'topic-centred'; that is, it has been used as a tool to investigate other topics of interest. He suggests that there has been less 'theory-centred' research, where the primary interest is in the theory itself and its development. Like Cramer, he suggests that further research needs to be done to test hypotheses generated by the theory. Apter suggests a number of areas where knowledge and research are currently incomplete. Among these are: reversals between pairs of transactional states; satiation (a proposed condition for reversal); transactional emotions, where there is a need to verify the emotions related to transactional states; that reversals do indeed result in the changes in emotion predicted by the theory; more focus on state (as opposed to dominance); and patterns of behaviour over time, including repeating sequences of states or state combinations over time.

However, the position taken in this study is that Reversal Theory at the moment is an adequately robust theory, which is well evidenced and has proven research utility, but that is subject to challenge and amendment over time. In the meantime, the theory provides the most comprehensive and coherent description of motivation and emotion, and will provide a theoretical underpinning and interpretative frame for this study. One of the consequences of using the theory is the primacy of the construct of motivation within the field of affect, and this study will seek to characterise disaffection, and related affective constructs, as primarily a motivational phenomenon.

Emotions and Reversal Theory

One of the advantages of Reversal Theory, and viewed specifically from the perspective of the current study, is that it gives an account of the primary emotions, and relates them to motivational states and features of our experience of the world. This is a significant advantage in that both theoretically and empirically, it is clear that motivation and emotion need to be acknowledged and taken account of in a full characterisation of disaffection. Here, we will give a brief account of emotions from a Reversal Theory point of view. A fuller account can be found in Apter (2001).

There is a supra-organisation in the eight states of Reversal Theory. This is based on the fact that there are two primary variables that control our emotional lives.

One is the experience of arousal, and this relates to the means-ends and the rules domains, and this is why those four states are called the somatic states. The other variable is felt transactional outcome (win or lose), and it relates to the transactional states of mastery-sympathy and self-other.

First, the somatic states. It has long been recognised that there is a pattern in the relationship between felt arousal and the psychological phenomena of hedonic tone – in effect how positively or negatively we experience arousal. The traditional approach (still current in many textbooks) is optimal arousal theory. This illustrates the relationship as an inverted U-curve. There are a number of problems with the notion of optimal arousal, but not least that it is difficult to accommodate a point in the space for pleasant high arousal. Reversal Theory gets over this problem by positing the relationship as in Figure 1. The validity of this interpretation has been supported empirically by substantial psychophysiological evidence.

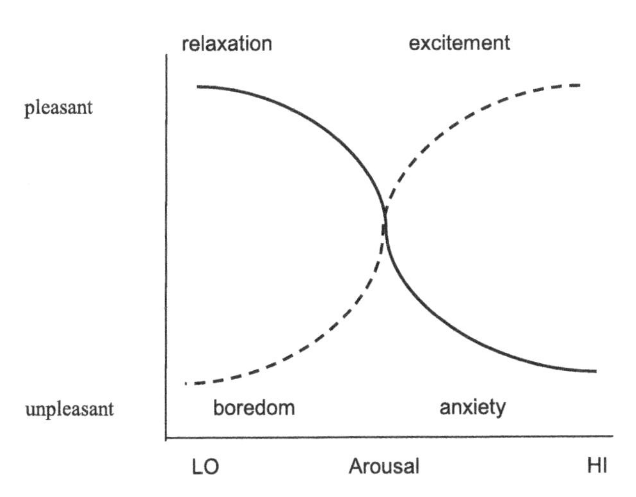

Figure 1. Conforming somatic emotions

Figure 1 shows the relationship of arousal to hedonic tone for the serious and playful states when combined with the conforming state (i.e. serious + conforming and playful + conforming). Although a single label has been chosen for each corner of the space, in fact, each point more realistically corresponds to a bucket of closely related and finely nuanced emotions. So, for instance, anxiety could also manifest as panic, fear, or even terror, if extremely intense. There is an equivalent relationship between the serious and playful states when combined with negativism, which also produces four emotions. They are listed in Table 4.

There are many subtleties and complexities arising from this model, but one interesting fact is that in the serious state we will be arousal avoiding (because high arousal is experienced as negative (anxiety). One way of increasing arousal is to put oneself at risk in some physical or psychological sense. So, when in the serious state, we will be averse to risk. Conversely, in the playful state, we will be arousal seeking, and may actively seek out danger or risk. This observation has relevance for students undertaking problem-type activities in mathematics classrooms.

In the mathematics education context, one of the key concepts defined and studied is maths anxiety. The Reversal Theory framework enables us to take a systematic look at the relevance of boredom and excitement seeking, as well as curiosity, which is a phenomenon not covered in any substantive way in the literature.

There is a similar pattern of emotions in relation to the mastery-sympathy and self-other pairs of states in relation to the other key variable of felt transactional outcome. In simple terms, felt transactional outcome is winning or losing. Combining these states, there are two comparative 'butterfly diagrams, (as in Figure 1), for the transactional states. The state combinations, and their related emotions are shown below.

Table 4. Emotions and states

State combination	*Positive emotion*	*Negative emotion*
Serious-conforming	tranquillity/relaxation	anxiety
Playful-conforming	excitement	boredom
Serious- rebellious	placidity	anger
Playful-rebellious	mischievous	sullenness
Self-mastery	pride	humiliation
Self-sympathy	gratitude	resentment
Other-mastery	modesty	shame
Other-sympathy	virtue	guilt

In Reversal Theory, emotions are evaluated in terms of hedonic tone. In this way, emotions are seen to be experienced as either pleasant or unpleasant. Anger, for instance, is experienced as unpleasant.

However, despite the advantages of the account of emotions as set out here, a critical perspective is also needed. Whilst there is evidence that supports both the existence of emotions related to particular states, and that reversals are associated with the emotions predicted by the theory in relation to the somatic states (Apter 2001), there is a scarcity of such empirical evidence in relation to the transactional states. It follows that conclusions related to transactional emotions may need to be guarded, and reliability will be borne in mind in discussions, and referred to in the evaluation of the frameworks and instruments used.

In this study, the theory is used in a number of ways. It will be used as a design framework to inform the development and deployment of methods and instruments. Secondly, it will be used as an interpretative framework for use in analysing interview narratives. It was hypothesised that evidence of states, of their related emotions, and of reversals themselves will be evident in young peoples' accounts of their experience of learning (or not learning) mathematics. Finally, it is expected that by relating the theory to the empirical data it will be possible to develop newer, formative theoretical constructs in the field.

Before that, however, it is possible to offer a critique of other theories from a Reversal Theory perspective.

CRITIQUE OF SELF DETERMINATION THEORY[2]

Problems with Intrinsic and Extrinsic Motivation

Despite the influence of SDT, the theory is not without dispute and indeed, even controversy. For instance, there are questions over whether, or to what degree, rewards have a negative effect on intrinsic motivation. Over the years, Deci and Ryan and others repeated the experiments, and conducted extensive research on the notion of rewards. However, the results were not always replicated, and so the theory built on the results, (self-determination theory) also came into question. A meta-analysis, (Cameron & Pierce, 1994) concluded that the undermining effect of extrinsic rewards on intrinsic motivation was minimal and largely inconsequential. A vigorous debate pursued in Review of Educational Research, questioning the methodology of the meta-analysis, and this resulted in Deci and Ryan responding with their own meta-analysis (Deci et al., 2001b) which confirmed such an effect.

It is not the point here to join either side of this debate, except to say that the empirical evidence is not unequivocal. This leads to the question of are there possible explanations for the confusing or null results. In addition to that, the generalisability of the theoretical position of SDT is also in question. Hidi and Harackiewicz (2000) have pointed out that Deci and Ryan's own meta-analysis investigated the effects on relatively short term and simple tasks. They further point out that it may be inappropriate to assume the same relationship applies in relation to long-term or more complex tasks. They also point out that studies that included uninteresting tasks were excluded from the meta-analysis. Where they were taken into account, rewards did not reduce intrinsic motivation.

One of the key benefits of SDT is that it provides a qualitative elaboration of motivation, and this is perhaps why it has been so influential. So, for instance, Hannula, along with many others, adopts the core needs of autonomy, competence and relatedness as the primary human motivational needs. However, as has been pointed out, the theory does not necessarily account consistently for all of the known data. It therefore makes sense to identify and evaluate an alternative theoretical framework. Using Reversal Theory it is possible to critique other theories of motivation, and to identify and address anomalies and omissions in those theories.

The notions of intrinsic and extrinsic motivation have gained widespread traction in the literature on motivation in educational research. Indeed, in much reported research, it is taken as a given. It has had a powerful influence on theorising in the field, and indeed into policy and practice. Self determination theory (SDT), a theory derived from the notions of intrinsic and extrinsic motivation, has particular relevance for mathematics education. It gives an account of the potential damage that externally imposed requirements, regulation, rewards and sanctions can have on motivation, and because, on the contrary, it stresses the need to engage students' natural interest and curiosity, it has provided an influential theoretical basis for policy and practice in mathematics education (Deci et al., 2001a; Hidi & Harackiewicz, 2000; Niemiec & Ryan, 2009). In addition, it appears to be highly relevant in providing some explanatory basis for phenomena reported in studies such as those of Boaler (2009) and Nardi and Steward (2003).

In this section, I will examine the nature and the claims of SDT and its constructs, and seek to identify and clarify ambiguities in the theory.

Using ideas from Reversal Theory, there are a number of challenges that can be offered to SDT as it stands. By examining these challenges from a Reversal Theory perspective, new light can be shed on some of the problematic areas and reported phenomena.

Pink (2010) gives a good account of the history of the experiments that examined the effects of rewards and punishment on behaviour, at first on monkeys, and later on students (an incremental improvement?). The series of experiments and the body of knowledge appeared to show that rewards can kill motivation, but, on the other hand, humans have an inherent capacity to seek out novelty and challenge, and to extend and exercise their capacities. Pink himself cites the creation of not-paid-for or rewarded volunteer projects like the creation of Linux and of Wikipedia as evidence of this intrinsic motivation. A more recent but just as interesting phenomenon is that of the so-called 'gamesmakers' of the 2012 London Olympics. Just what inspired and motivated these extraordinary people to do what they did, with such obvious pleasure and commitment, is not easy to explain in terms of many current theorisations of motivation. One thing that must surely be true is that the motivation must be multiple and many-layered to sustain processes of application, selection, scheduling and their subsequent duties and responsibilities over many weeks.

One way in which SDT is ambiguous is in relation to the notion of reward itself. The notion of reward is inherited from the behaviourist tradition. In behaviourist experiments, food for rats, and tokens for students are deemed to have an obvious and stable value. However, the subtle point here is that any external moderator is context-sensitive. Indeed, the meaning of the same moderator can be interpreted in different ways according to the motivational state of the perceiver. Take food as an example. It can mean different things at different times. Using the eight states of Reversal Theory, it is not enough to say that one is motivated to eat food. It is necessary to say *how* one is motivated to eat, and that depends on the motivational state. For instance, in the serious state, one might be motivated to eat to satisfy hunger. On the contrary, in the playful state one might be motivated to eat in a restaurant for the sensual pleasure of eating the beautifully prepared food. One might attend a family Sunday lunch in the conforming state, to please our parents for instance. In the rebellious state, one might eat precisely because it is illicit, if we are fed up being on a diet and we attack the chocolate biscuits. We can express mastery by developing and using our skill in cooking. By eating healthy food, we might experience sympathy for self, or a sense of self-nurturing. By enjoying company, sharing, and the social aspects of eating, we will be motivated by the other-oriented sympathy state. Thus food can at various times and in different ways, satisfy all eight motivational states. Because of this, it is potentially a motivationally rich activity, which is to say that whatever motivational state we are in, we can gain the satisfaction of the state through food. So because different motivational states give different meanings to life, food (or indeed most things in life) does not have a single consistent motivational value. The same will be true for significant aspects of our educational life and experience such as mathematics.

One implication of this is that rewards may not work in experimental conditions as expected because they are not perceived in the way that researchers assume they are. This alone may explain some of the confusing results of experiments reported.

Another problem in the body of SDT research is that different definitions for intrinsic and extrinsic motivation are used, even sometimes by the same author at different times, and this creates confusion. One class of definitions for intrinsic motivation take it to mean doing some activity for its own sake, and of extrinsic motivation meaning doing it as something instrumental to achieving something else. However, some researchers take intrinsic to mean that one's behaviour is determined by oneself, whilst extrinsic motivation is determined by the environment. These latter can be termed 'locus of control' definitions.

But these different definitions involve quite different dimensions. For example, the environment could determine that one does something for its own sake (in the playground, on the tennis court, etc.). This would be extrinsic by the locus of control definition, but intrinsic by the 'ownsake' definition. Alternatively one could decide for oneself to do something where there were external rewards to work for (for example, to decide in the face of family opposition to become a musician). This would be intrinsic by the locus of control definition, because one is making one's own decisions, but extrinsic by the 'instrumental' definition since the hope is that it will eventually provide external rewards like playing in an orchestra and earning money, and one is willing to work towards these. So the two definitions do not necessarily go together. Indeed, they are quite incompatible. You can do fun things that you have been told to do, and serious things that are self-chosen.

Various other definitions have also been given. For example, Purkey and Schmidt (1987) defined intrinsic motivation as the motivation to engage in activities that are designed to enhance a person's self-concept. In this case a whole host of activities that are both self-chosen and pleasurable in themselves would have to be excluded, and would therefore fit other definitions of intrinsic, but do not appear to be about enhancing one's self-concept, like getting drunk, taking drugs, engaging in illicit sex, and so on.

In Reversal Theory terms, the 'own sake' definitions fit very neatly with the serious-playful distinction. This also gives insight into why intrinsic motivation (under this definition) is damaged by rewards, because quite simply, the nature of the task has been re-defined from play to work. In the serious state, the outcome will be in focus (whether achieving the task or gaining the reward), and pleasure in the activity itself will not be a feature of the phenomenal frame at that moment.

But it is also interesting to read descriptions of intrinsic motivation. Niemiec and Ryan (2009) after Ryan and Deci (2000) define them thus:

> Intrinsic motivation refers to behaviours done in the absence of external impetus that are inherently interesting and enjoyable. For example, when people are intrinsically motivated to play, explore and engage in activities for the inherent fun, challenge and excitement of doing so…accompanied by feelings of curiosity and interest. (p. 134)

These notions of interest, enjoyment in the task, fun, challenge and so on, make it sound exactly like paratelic enjoyment.

The locus of control definitions are consistent with the conforming/rebellious states within the domain of rules. That is: I either do something because I am required by the environment to do it, or I do it of my own volition, to assert my individuality, even in the face of disapproval. This thus may address some issues of confusion, because it is perfectly possible to combine these motivations in four ways:

- Serious-conforming (A) – I can be goal-focussed in relation to a goal (external outcome) required of me by the environment. E.g., my parents want me to do well in my maths homework
- Serious-rebellious (B) – I can be goal focussed on a goal chosen by me in the face of opposition. E.g., I'm going to be an actor, even though my parents want me to be an accountant
- Playful – conforming (C) – I can please myself within a defined context and set of expectations. E.g., Getting immersed in the momentary pleasure of solving a problem in a mathematics classroom

- Playful-rebellious (D) – enjoying myself in the moment when I am not supposed to. E.g., using a coordinate worksheet in a maths classroom to make 'naughty' words

Under the 'own sake' definition, C & D would be intrinsic. Under the locus of control definition, B & D would be intrinsic.

The introduction of the requirement that intrinsic motivation has to satisfy the basic needs of competence, autonomy or relatedness may contaminate the notion of doing something purely for fun or enjoyment. Isn't doing something for its own sake exactly the opposite of doing it for a separable outcome? It changes the requirement from play to work. Doing something to increase my competence may not be enjoyable in itself. Learning scales on a piano is an example. Asserting our autonomy (for instance, challenging or confronting our parents) may be extremely uncomfortable, and not something I am doing for the pleasure of it. (Although, clearly, there are times when young people do this purely for the pleasure!)

At the same time, doing things purely for the enjoyment does not have to involve competence, autonomy or relatedness. There are activities that people do for their own pleasure that do not increase the sense of competence, autonomy or relatedness, such as getting drunk.

In Reversal Theory terms, the key distinction highlighted in the paradox of the original experiments is that between work and play. In the playful (paratelic) motivational state, outcomes and consequences are ignored and we behave in a spontaneous and risk-seeking way. It seems that, just calling something work or play can make all the difference.

Hidi and Harackiewicz (2010) report the work of Sansome and colleagues who have shown that students can use strategies to make boring tasks more interesting, such as making a game out of them, particularly when provided a reason to value the activity.

The argument above is also not meant in any way to deny that competence, autonomy and relatedness aren't important aspects of motivation. However, in addition to the ambiguities discussed, empirical evidence in support of Reversal Theory suggests that there are eight and not only three motivational states. In this way, SDT may indicate important aspects of motivation, but is incomplete as it does not take into account these additional aspects of our motivation. SDT gives no account of negativistic behaviour, for instance. Nor is altruism or self-sympathy (the need for love) easy to compute in terms of the SDT as it stands.

THE STRUCTURE OF AFFECT

Using the notion of the centrality of motivation and emotion, it is now possible to set out the way in which affect is conceived in relation to disaffection. This schematic represents the understanding of the structure of affect that is derived from the analysis above, and is used as a basic set of assumptions in the current study.

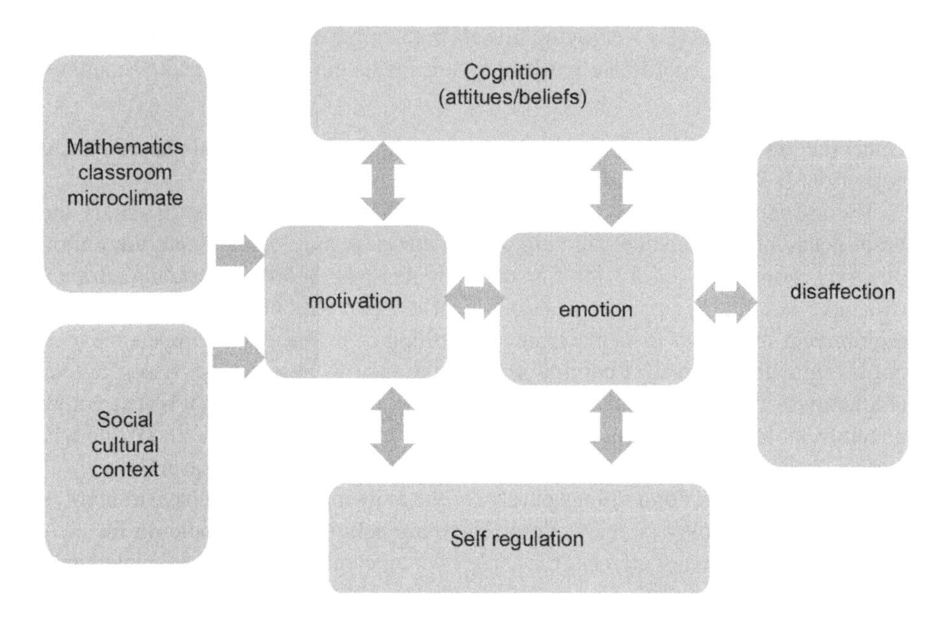

Figure 2. The structure of affect with respect to disaffection

A number of features in the diagram should be remarked upon: motivation is seen as central to our understanding of the notion of affect. Alongside motivation, emotion is also highly important in the experience of learning (or not learning) mathematics, and intimately connected to motivational state, as outlined above. These two constructs are viewed as the core of our phenomenal experience from which all else ultimately derives. Reversal Theory gives us a means of characterising relationships between motivation and emotion, and the mechanisms by which emotions arise.

Learning as a social text (Hannula), and the 'social turn' (Lerman) needs to be accounted for in the schema. The multiple levels of socio-cultural context (race, ethnicity, citizenship, gender, social class, classroom climate, etc.) are influences on the learner and the learning process. In particular, the context of the classroom itself is shown as of equal weight with socio-cultural context in general. However, these influences are not privileged over any other influence or system.

This is also related to another way of categorising these systems: Since we have proposed that motivations and emotions are the core of our subjective experience, alongside cognition, we can use the term (and the metaphor) infrastructure. It seems reasonable to assume that we are born with these systems as capabilities, although they become populated by experience, and are subject to development over a lifetime. The more cognitively mediated constructs such as attitudes and beliefs, we can label 'superstructure', and develop with accumulated experience.

NOTES

[1] Although Svebak (1993) is one such study.

[2] I am grateful to Dr Michael Apter whose writings brought some of these ideas to my attention, and with whom I have discussed these issues at length.

CHAPTER 4

THE STUDIES

A number of themes and conclusions have emerged from the review above which influences the nature and scope of this study. Firstly, disaffection has been located within the field of affect, and a structure for the affect space has been proposed. The review has identified that there is a need to move away from the purely quantitative study of strongly cognitive constructs like attitude and beliefs. It is proposed that motivation is central to the affect space, and alongside this, the concomitant emotions have an important part to play in our subjective experience, and this suggests that an investigation of disaffection with mathematics needs to centre on these notions. It has also emerged that there is a need to counter the study of populations in order to study individuals more closely, and to understand the subjective experience of disaffection more than is currently the case. This locates the study in the investigative or interpretive frame, and in addition, it is seen as desirable to look for new theoretical frameworks and methods with which to research the problem.

The aim of this research then, is to investigate the nature of students' experience of disaffection with school mathematics, and the factors that contribute to it. A number of sub-questions will be explored:

- What is the incidence and occurrence of disaffection within the population of young people from 14–19 years old?
- How is disaffection with school mathematics experienced and expressed by students?
- What are the motivational and emotional factors that contribute to disaffection with school mathematics?
- What factors influence how young people learn (or do not learn) mathematics?
- How does students' relationship to mathematics change over time, and why does it change?
- How can Reversal Theory be used as a theoretical framework to capture the richness and complexity of student's experience of school mathematics, and to enhance our understanding of disaffection?

In order to answer these questions, a range of qualitative and quantitative methods will be developed, through a series of preliminary studies, to elicit data from the perspective of a wide range of young students who are or may be disaffected with school mathematics. These methods will serve the main study, which is centred on one-to-one interviews with disaffected students.

Given these considerations, the approach adopted in this study is a constructivist and interpretivist one, which has the intention of understanding the 'world of human experience' (Cohen et al., 2000) in which the participants' view of the world is privileged.

However, it is claimed:

> Very few studies of learning which adopt a participatory perspective deal adequately with individual learner. (Hodkinson & Macleod, 2010; Hodkinson & McLeod, 2011, p. 178)

Hodkinson and McLeod endorse a constructivist approach, claiming "the construction metaphor centre upon the ways in which people make sense of any learning experiences they have" (p. 179). They note some of the advantages of a construction approach to learning, and the study of life history as a means of understanding the subjective reality of participants. They stress the efficacy of narrative construction through reflection. "There is an affinity between life history and conceptualising learning as a form of construction" (p. 179). They note that in particular, a study of life history can be very effective in bringing out the practical, physical and emotional dimensions of learning that are often omitted in more cognitive accounts. These distinctive features of the constructivist account of learning suggest a strong consistency with Reversal Theory in that they are both rooted in subjective reality, and both focus on the psychological and emotional aspects of life. Consequently a mixed methods approach is adopted which provides a richer range of data and is "a key element in the improvement of social science, including education research" (Gorard et al., 2011), as quoted in Mackenzie and Knipe (2006).

Since the primary area of this study has been declared to be motivation and emotion, it has been argued that Reversal Theory is seen as the most general and comprehensive theory that covers these areas. It will therefore be used as the primary theoretical approach, notwithstanding that other theories will be referenced as appropriate.

Notwithstanding the methodological need for innovative instruments of data collection, the use of interviews is still seen as the primary means of interacting with students. Interviews are seen as a fundamental instrument in qualitative research (Cohen et al., 2000). In particular, when trying to understand the world of young people, they are seen to have value both for researcher and participant. For the participant they can have an empowering effect (Rudduck, 2002). Since the proposed interviews are seeking complex data that involves elements of life history, motivation and emotions, and can thus be highly individual and idiosyncratic, semi-structured interviews are appropriate. According to Cohen et al. (2000), they provide the researcher with the appropriate level of flexibility in doing this.

In order to explore and verify ideas and approaches, I conducted a number of studies preliminary to and parallel to the main qualitative study.

The Preliminary Study

In this study, and in order to provide original data for the study, seven young people were interviewed who were attending sessions and meetings at a local Connexions office in the UK. Because of the circumstances, it was likely that some of these students would exhibit aspects of disaffection, and that proved to be the case. These interviews provided very rich sources, expressing motivation, disaffection and emotion in authentic narratives.

I was able to negotiate access to the Connexions office in the East Midlands for a period of one day. Connexions is a service that helps, advises and supports school leavers, particularly those who are in danger of becoming NEET (not in education, employment or training). Since most of the young people who use this service have performed poorly in school, it was reasoned that many of them would have a level of disaffection either with school/education in general or with mathematics in particular. However, there was no way to make any assessment or evaluation to confirm this.

Although a position has been adopted that motivation is the foundation of affective experience, a case has also been made that other affective constructs such as emotions, attitudes and beliefs, also have an effect on engagement (or not) with mathematics, and with learning. Thus issues of liking or not liking, of interest (or lack of it), expressions of engagement (or lack of it), of attraction (or its opposite), will all be considered relevant to young people's relationship with mathematics. Some examples are cited below. These segments were then organised into categories which emerged iteratively and inductively (Thomas, 2009). Many of these categories were easily relatable to the Reversal Theory framework. Some themes emerged, however, which stood on their own aside from this framework in that they did not relate in any simple or direct way to motivational states or emotions, as specified within the theory.

At this stage it is worth presenting briefly some results and conclusions from this study, since they were influential in informing the development and choice of methods and instruments for the main study. Here are some salient points:

- Although all of the interviewees could be considered as disaffected, this disaffection manifested in very different ways for different students. Scott's narrative, along with his language and voice tone, all confirmed his over-riding negativity in relation to school mathematics. He mentions the word 'struggle' 17 times. By contrast, Sam sees mathematics as an intriguing mystery, but he feels excluded from the elite, and his ability to do mathematics (apart from basic calculations) does not match his philosophical perspective.
- The range of emotions reported includes: pride; humiliation; panic; embarrassment; excitement; enjoyment/fun; intrigue; powerlessness; competitiveness; anger; pointlessness; sympathy; anxiety; boredom; triumph. The richness of these student's emotional responses to mathematics suggest not only distinctive individuality, but that these areas need further and more systematic exploration
- The perceived and experienced epistemologies and pedagogy of school mathematics plays an important part of the disaffection of these students. They often don't see the point of what they are doing, and the diet of worksheets and textbooks is clearly alienating. Braidon sums it up by saying: "Books don't fulfil your brain. Hands on is better." Mel and Kelly say: "Most of it is bookwork…boring."

Such observations and conclusions arise from a direct consideration of the affective aspects of the narrative and they stand alone in their own right. However, some aspects of the data are thrown into sharper relief by using Reversal Theory as an interpretative framework. To give a few examples from this exploratory study:

- It is very clear from these stories that performative self-mastery (being able to 'do') is not only massively important to students, but is also intimately tied up with understanding. This is typified by the statement from Mel and Kelly that "when you understand you can enjoy them – they become quite enjoyable."
- Self – sympathy plays an important role in these young people's experience of school mathematics, and in their disaffection with it. Feeling ignored or uncared for has a corrosive effect on their experiences, as does being labelled as 'dumb'. Scott's experience of 'no-one's helping me' is typical here.
- The importance of the paratelic (playful) state emerges strongly in these narratives. There is a good deal of evidence of the importance of fun. Exercises or approaches that move away from arid worksheets and textbooks are valued and add excitement. As Mel and Kelly put it: "teach it in chocolate bars." In the negative sense, the paratelic state is also evident in the multiple mentions of boredom, where the need for positive arousal is present but unsatisfied.

The results of this study were reported more fully at PME (Psychology in Mathematics Education conference) 35 (Lewis, 2011).

This exploratory study has achieved a number of things. Firstly, it has provided confirmation of aspects of the TIRED framework as identified by Nardi and Steward. Secondly, it provides an ethnographic map which portrays the motivational landscape of disaffection through the voice of disaffected students. Thirdly, the study's findings demonstrate the complexity, richness but also the volatility of motivation and emotion in mathematics classrooms which has enabled me to identify a number of key aspects of motivation that I will argue in the conclusion need further exploration in relation to disaffection, as outlined in the analysis.

The study has also provided further evidence of how motivational states are experienced and expressed in a mathematics education context, and has confirmed the utility of the Reversal Theory framework in mapping the motivational and emotional landscape of young disaffected students.

On the basis of the exploratory study, the following conclusions were reached:

- The use of a psychometric instrument would provide a useful addition to the repertoire, and particularly if it could address the experience of negative emotion in a coherent way. The use of such an instrument would allow for data collection from a larger sample than could be reached by interview alone, but would also provide data to discuss within the interviews.
- That prior experiences and life history in relation to school mathematics was a rich potential source of data, and so the collection of data in relation to students'

past relationship to mathematics would be extremely useful if it could be collected in a systematic but simple way.
- The study showed that positive experiences are present and relevant to students' learning of mathematics, and a structured way of accessing such experiences is needed.

These points were then reflected in the revised data collection protocol that was subsequently developed.

THE MAIN STUDY – METHODS AND INSTRUMENTS

As a result of re-thinking and re-designing the interview protocol on the basis of the exploratory study, the following methods have been used in this study.

TESI-ME

The exploratory study had highlighted a whole range of negative emotions that were associated with the learning of mathematics by the students in that study. This had alerted me to the need to collect more systematic and structured data about the experience of negative emotions, and a survey instrument seemed to be an appropriate way to do this. The purpose of the questionnaire in this study was to gauge the prevalence of negative moods and emotions in a group of students who were likely to be disaffected. It was seen as a way of characterising disaffection through emotion rather than attitude. In effect it provides answers to the questions; *'how stressed are you about mathematics?'* and *'how do you experience that stress?'*

The Tension and Effort Stress Inventory (TESI) (Svebak, 1993), was designed to be a one-page survey measure with an integrative orientation to the experience of stress. It is based on the Reversal Theory account of unpleasant emotions or moods (which is why, in this case, only the eight negative emotions are addressed, ignoring the eight positive emotions specified in the theory). "The TESI has proved to be a practical instrument for quantitative assessments of the subjective experience of exposure to stressors" (Svebak, 1993, p. 202). It can be used on a face-value basis without recourse to the theoretical assumptions inherent in its design. It is based not on behavioural or biological bases of stress, but about the subjective experience of stress. The eight emotions are: boredom, anxiety, anger, sullenness, humiliation, shame, resentment, guilt. It is reasoned that disaffection will be associated with the experience of such unpleasant emotions.

The test was adjusted to make the wording appropriate and relevant to the mathematics education context (TESI-ME). In addition, the labels for the emotions were altered where necessary to reflect current language use within this social grouping. Students were asked to report the degree to which, in relation to mathematics, they experienced stress, effort, and the eight negative emotions on a Likert-type scale from 1 to 7.

Tension & Effort in Mathematics Inventory (TEMI)

Code	Name	Gender	Age
SW2 - A - 6			

Please rate the foll[illegible] the appropriate number

A How much pressure, **stress** challenge or demand are you experiencing in doing mathematics at College

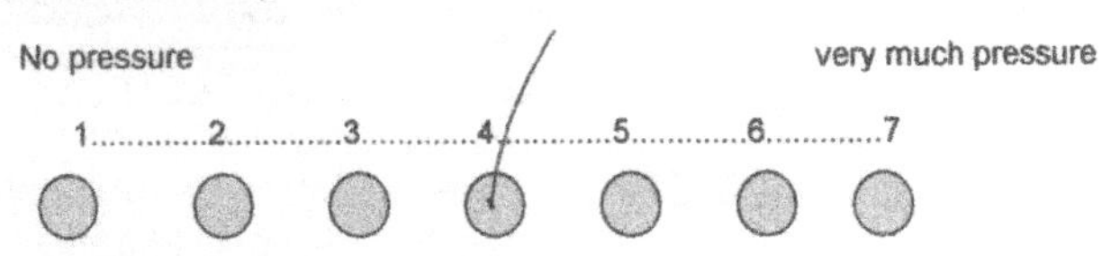

B How much **effort** do you put in to the current situation to cope with the pressure of doing mathematics in College

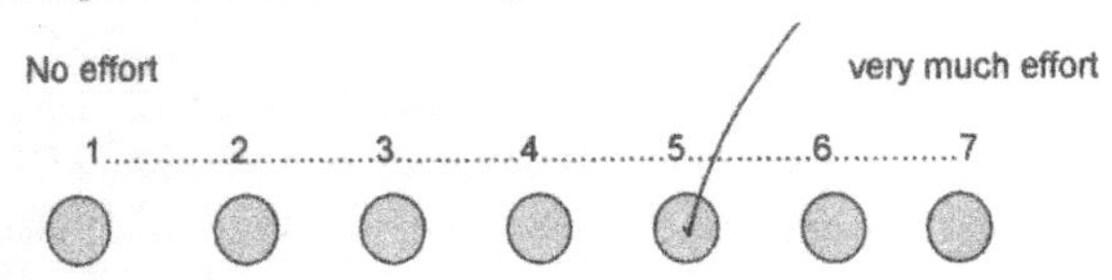

C How much do you experience the following **moods or emotions** when doing mathematics in College

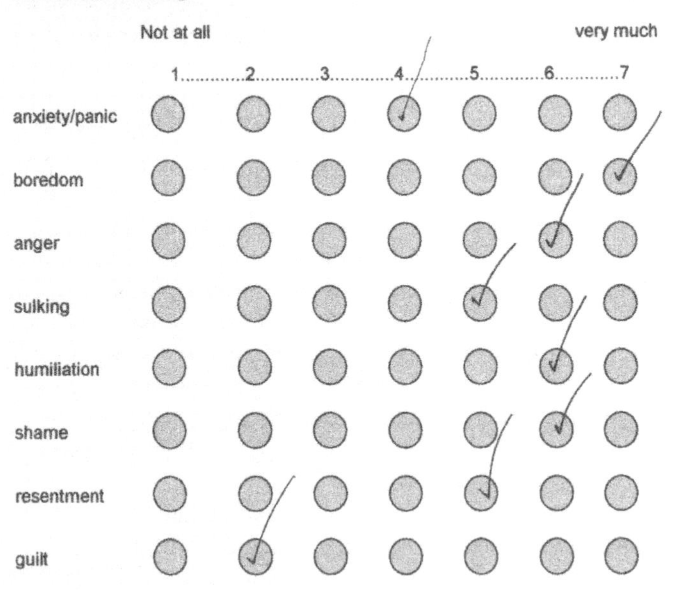

Figure 3. A completed TESI-ME
(Adapted from Svebak, 1993)

In the first instance, the TESI-ME was administered by the author to each of the participating classes in two colleges. This data, together with class observations, and triangulated with the judgements of teacher/lecturers, was used to identify students who were likely to be disaffected with mathematics. Because of the vagaries of attendance in this population, and the fact that students were only interviewed if they volunteered, the sample can be regarded as an opportunity sample. The TESI-ME completed by each student was available at interview, and formed part of the discussion.

Life Histories

During the interview, students were invited to plot their relationship with mathematics over time. This instrument, 'Me and mathematics', was a grid with a horizontal axis marked out with school years 1 through 12, and a vertical axis marked from –5 to +5. Students were invited to place a point on the scale for each school year. In this way, they plotted a graphical representation of their mathematics life history. This was then discussed in the interview. This simple instrument enables the exploration of the question; *'How has your relationship to mathematics changed over the years'*, and subsequently, *'What causes rises and falls in affect?'*

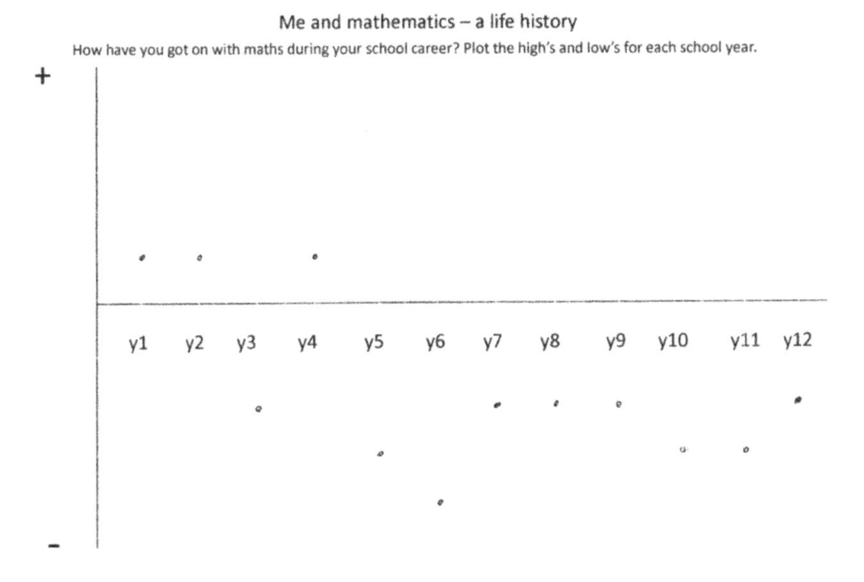

Figure 4. A completed life history

Such visual methods are increasingly being used in educational research, as they seem to be particularly useful for younger, less articulate or marginalised groups (Davidson et al., 2009). Chula (1998) has investigated the use of drawings as a methodological technique for visual data analysis in the study of the perceptions of adolescents' experience of education. She concluded that drawings are useful as a singular source of interpretive inquiry, and she makes the case for drawings in research as an alternative, non-discursive form of knowing. Chula sets out a number of purposes that the use of drawings can satisfy. These include: as a stimulus to the retrieval of thoughts; as a means of expression and articulation where words are difficult to find; as an interface between the interviewer and the individuals. She also points out that a further advantage is that new and emerging theories of learning have implied a range of learning styles, and that the techniques under discussion can offer a means of expression for people with more visual styles of learning.

Borthwick (2011) also uses visual techniques in her investigation of young pupils' views in relation to their experience of school mathematics. She points out that visual representations can be used to elicit expressions and emotions for pupils for whom verbal expression may not be easy. Davidson et al. (2009) also argue the case for the use and acceptance of visual sources in qualitative research. Their study encompasses a number of features that characterise the most important affordances of using visual sources. One of these is the access they give to complex realities and multi-layered meanings in the subjective experience of those studied. That these methods are able to access meanings, interpretations and themes not possible through other methods is also reported (Sewell, 2011). Davidson goes on to talk of "the hidden consciousness of their experience", and states that "Visual images have the capability of bypassing cognitive defences of our experience to tap directly into our emotional and spiritual/ intuitive zones of consciousness" (Davidson et al., 2009, p. 10).

However, a number of commentators also point out that such visual data can be reinforced by combining it with other means such as interviews. For instance, Chula states "When interpreting beyond what is visible and descriptive, other methodological techniques such as written narrative and interviews are necessary to clarify ambiguity and vague symbols, and to maintain the integrity of the stories told" (Chula, 1998, p. 1). This is echoed by Croghan et al. (2008) who say "combining verbal and visual forms of self-presentation allows individuals more scope for presenting complex, ambiguous and contradictory versions of the self" (Croghan et al., 2008, p. 355). Chula also points out the importance of the interview alongside the drawing. This not only allows for the elicitation of meaning within the drawings, but allows for the elicitation of data beyond the drawing itself.

The life history was produced during the interview. The participant was offered a blank template to complete. The diagram was explained. They were asked to recall year 1 and to rate how positive or negative their experience of school mathematics was. They were invited to plot this as a point on the graph. They were then invited to make the same judgement for each year of school, up to the present. Discussion proceeds from here on the basis of comments or questions about key patterns or features by the interviewer. Such questions were of the type: *'That looks quite positive in the early years.' 'There is a large dip in year 3. What happened there?'*

Care needs to be taken in interpreting this data, and in assessing its trustworthiness. Since the data will not be aggregated, and no inferences will be drawn from the life histories as a body of data, it can be interpreted entirely in an individual way. That is, the focus, within the interview, and in reporting the results will be confined to the subjective meaning for the individual concerned. No judgement will be made as to the objective or truth status of the data. Whether a rating of +5 by participant X, represents the same affective reality as a +5 by participant Y, will not be of concern here. It will be assumed that there is some internal subjective consistency of meaning in that the student's memory of a year rated +5, will have a different affective meaning than one rated –5, and that rises and falls in perceived affective quality (particularly if they are dramatic) will be meaningful.

Cards

During the course of the interview, students were offered two sets of coloured cards in turn and asked to choose those cards which represented something meaningful about their own experience of mathematics. The first set of cards represented positive emotions, as indicated by Reversal Theory. The emotions included were:

> excitement/curiosity, relaxed, mischievous/playful, proud, virtuous, modest, grateful.

In the second set, each card was labelled with a motivationally significant or valent word or phrase, again all positive. Again, these were suggested by the eight motivational states as defined in Reversal Theory. They included:

> feeling cared for, sense of achievement, freedom, sense of purpose, helping others, powerful/in control, feeling part of the group, having fun, sense of duty.

Every interviewee voluntarily chose at least one card from each set. The card sort was an attempt to understand the presence of more positive affect in the experience of these students. In effect, it answers the questions: *'In what ways do you feel positive emotion, what motivates you, and where do you derive satisfaction?'*

The card sort idea was based on the original idea of the Q-sort, developed by William Stephenson in the 1930's, the use of techniques of selecting/sorting prompts presented on cards has been used and developed over the years in psychological and sociological research. Indeed there has been a substantial number of papers published referring to the technique (Thomas & Watson, 2002). They comment that the method offers "a powerful, theoretically grounded, and qualitative tool for examining opinions and attitudes" (Thomas & Watson, 2002, p. 141). Although the original method was structured by having participants rank statements on cards, thus producing quantitative analyses, the method used in this study was modified by simply asking participants to choose cards that represented their own experiences. These choices could then be discussed at interview.

The Interview Protocol

The methods outlined above were all either available for discussion during the interview, or generated within it. The interviews then explored the rationale for their choices and involved probing for further meaning and specific examples. Further questions were asked to elicit aspects of their positive and negative experiences of classroom mathematics. The interviews followed the following structure and flow:

- Introduction and rubric
- Life history (produced during the interview)
- TESI-ME – key patterns explored
- Card sort – choices discussed
- Experience of school mathematics
    - How good do you feel you are? If you are not successful, why not? What would it take to be successful? What makes other people successful?
- The mathematics classroom
    - What do you like about maths? What do you not like? Best and worst experiences? 'It would be better if……?

Figure 5. Interview flow

Notes were taken, and these were filed along with the relevant TESI-ME and life history diagram. The interviews were audio-recorded and then transcribed, and the data was entered into Nvivo. Passages were then coded to highlight motivationally and emotionally significant passages, events and interpretations.

The Participants

The exploratory study had alerted me to the idea that disaffection is perhaps at its most prevalent and acute in the year of national examinations at aged 16 years, and immediately afterwards, and it made sense that this was where it was perhaps easiest to find students who were indeed disaffected with school mathematics. One of the advantages of working in a university department of education is that the full time staff are very well connected to the local educational institutions. For a postgraduate researcher, access is perhaps the most precious resource. My colleague Sue Forsythe, who was the secondary PGCE tutor, introduced me to J and M who were tutors at Further Education colleges in the region. I introduced myself to each of them. I outlined my interests and suggested programme, without, at that time, a very exact plan as to who I might work with, or indeed, how, to some extent.

Both J and M were tutors and course leaders on a Use of Mathematics programme. Students who came to college having not gained a grade A to C in their GCSE exams, but who wished to study any subject at college, were required to take this mathematics course, either at foundation level or at Level 1, depending on their results. Both J at College R, and M at college G had four classes each on the level 1 programme. There was a difference in that at college R, each of the four classes had different tutors, whereas at college G, M taught four of the classes herself. The attraction of this context for this study is that since the students on these courses had 'failed' at national examinations at aged 16, and were more conscripts than volunteers, a significant number of them were likely to be disaffected with mathematics.

Both colleges were similar in their profile, in that they served catchments that are urban and modest-to-poor in socio-economic terms. A large proportion of students were from ethnic minority backgrounds, with English as a second language. Both J and M had agreed that I could come into the four classes and administer the TESI-ME questionnaire. From there, individuals would be invited to volunteer to be interviewed by me.

I felt it was extremely important that I administered the questionnaire myself. This was for a number of reasons. British Psychological Society (BPS) guidelines on the use of psychometric tests that they should be administered in standardised conditions to support reliability. Secondly, I felt that it was only me that could present my interests fully and accurately. In addition, I wanted to be able to set the tone in the right manner. By this I mean that I was able to emphasise that the questionnaire was not a test or examination, that it was voluntary, and that I viewed the responses as a favour. Also, I was able to emphasise that I valued honesty, and I was comfortable for them not to complete it rather than complete it with inauthentic data. I was also able to ask if they would be willing to volunteer to be interviewed, and I (hope) I was able to make clear that this would be non-threatening. It should be emphasised that the TESI-ME only takes a couple of minutes to complete. Another reason that it was important that I administered the test myself was that I was able to observe the students completing it. I was thus able to spot if someone was struggling with the language, and was able to respond to questions and requests for clarification.

This last point was significant. In the first version of the test that was used, great care had to taken about the label to use for the emotion 'sullenness', that would be as accurate as possible, but would be computable in the everyday language of the students. 'Can't be bothered' was the initial chosen label. Responses during administration, and during subsequent interviews led me to believe that this was being misinterpreted, and the label was changed to 'sulking'. In a slightly different way, in a number of classes students asked for an explanation of the emotion 'resentment'. In this case my view was that it was the subtlety of the emotion and not the label that was the problem.

In addition, I sought to develop a relationship with a school that had younger pupils across a fuller range of abilities. I was introduced to MG at a High School in the region. In this authority, High Schools take pupils in the range from year 6 to year 9. My interest was in year 9 pupils. I made 7 visits (being able to see a number of classes in each visit) in each of two school years – i.e., with different cohorts. Classes observed, and data acquired was weighted strongly towards the lower end of the achievement range (sets 4 and 5 out of five), since it was reasoned that there would be more disaffected students in those classes.

For completeness, it should be added that notes were also made either during or soon after observing lessons. I made notes of my conversations with both M from college G and MG from school G. I also interviewed both of these teachers, and another experienced teacher from college G. In addition, I wrote quite lengthy reflective statements detailing my thoughts about progress so far, and pondering decisions yet to be made.

Table 5. Details of preliminary and main studies

Study	*Description*	*Data sources*	*Methods*
Exploratory study	To gain primary data on the experience of disaffection	7 young adolescents	Semi-structured interviews
School K whole cohort study	Survey of a whole school cohort to provide contextual data on affect	208 pupils in year 9	Quantitative survey on aspects of affect and emotion
Main study		Year 9 pupils in school G, and students in College G and College R	Administration of TESI-ME to 184 students in Colleges G and R. Interviews with 48 pupils and students, using the instruments described above.

Analysing the Data

Data was analysed along the way, as and when it was collected in an iterative process. So, the first tranche of data was the data from College R and the first cohort of college G. The data was analysed and organised using a number of processes. For the first tranche I reduced the data by listening and transcribing 'significant motivational and emotional' passages. In addition, I collected the data derived from card choices so that I could aggregate all of the data against each topic (e.g., 'helping others').

At this point, it became clear that whilst all of the interviews generated interesting data, some were richer and more interesting than others. There were a number of threads that led to such a judgement, and I tried to be as thoughtful and explicit about this as I could. My first attempt was to attach to the 'transcript summary' a numerical value and a summary comment. It was quickly obvious that such a crude rating would be of little use. On reflection, the criteria I was using included: how disaffected the student was; how articulate they were; if the student was in some way different, individual or indeed even idiosyncratic; the 'vividness' of the story they told.

In order to analyse the data, each interview was fully transcribed. Initially, the analysis was performed on hard copy transcripts, but, over time, the transcripts were entered into Nvivo, and the coding was performed using the software. The data was coded by identifying passages that were 'motivationally or emotionally significant', which describes that issues of liking (or not liking) mathematics, of interest (or lack of it), expressions of engagement (or not), of attraction (or its opposite), and their associated emotions, will all be considered relevant. The style of coding, in the first instance, was to code passages using free nodes, and labelling these appropriately as they emerged. Passages could have multiple codings. As themes emerged across cases, and iteratively, these labels were gradually aligned.

Many such themes and labels were directly interpretable within the Reversal Theory framework. This is hardly surprising, since with the TESI-ME and the cards, these labels were 'designed in', as it were. In this way it was an explicit aim of the study to understand the experience of negative emotions such as anxiety and boredom, and the theory was used as a design framework in order to capture these. However, the analysis was not constrained by the requirement that passages had to be interpretable narrowly within the theoretical framework. Other themes emerged in a grounded way, and these are reported in more detail below.

One emerging conclusion was in relation to individuals. It became obvious that just telling the aggregate story would not do justice to the richness of the data. Individual stories were very different, and often highlighted the highly individual way that some students experience disaffection. It seemed the only way to properly and fully report these was as individual case studies.

Thomas (2009) points out that case studies have a strong history in educational research. A case study also provides the opportunity to apply a range of methods, so that triangulation of a variety of data from different sources can be combined to create a 'thick description'. In this way, a case study scores high on validity. However, one of the issues is reliability. Certainly, since it is known that aspects of socio-cultural context influence motivational and effective issues, there are issues of generalisability to address, and the basis on which the cases are selected should be scrutinised and justified in relation to the questions under research. But, as Cohen et al. (2000) also point out, this does not preclude certain specific kinds of generalisability. On the other hand, they point out that the case study is well suited to the interpretative-subjective approach. A case study can help to illuminate and report the complex, dynamic and unfolding interactions of events and relationships, including thoughts and feelings, since it blends description with analysis.

Nisbet and Watt (1984) (cited in Cohen, Manion, & Morrisson) describe a case study as:

> A specific instance that is frequently designed to illustrate a more general principle. (Cohen, Manion, & Morrisson, 2000, p. 181)

They also point out that it can help us to understand how ideas and abstract principles are effected in a real-life contexts. How this understanding is achieved can depend to a certain degree on the nature of the case study itself. Case studies can be exploratory, descriptive or explanatory (Cohen, Manion, & Morrisson, 2000). And although there are other classifications of case studies, all allow that an interpretative approach, fusing theory and practice is feasible. The cases in this study exemplify elements of both exploratory and descriptive studies. The cases were chosen as exemplary cases, since each illustrates different aspects of the experience of disaffection.

In order to better gain insight and to represent the full 'roundness' of the stories of these individuals, it was decided to adopt a more searching analytic technique. Following Smith and Osborn (2003) I adopted an Interpretative Phenomenological Analysis (IPA). The aim of IPA is to explore in detail how participants are making sense of their personal and social world, and the main currency for an IPA study is the meanings that particular experiences, events, and states hold for participants. It is phenomenological in that it is based on the lifeworld, personal experiences and perceptions of individuals, and it is hermeneutic in that it is concerned with the meanings and sense-making of those individuals. It is seen as a counter to the prevalence of quantitative and experimental approaches of much cognitive psychology, with their focus on nomothetic or aggregate picture of the world. It thus favours the qualitative and the idiographic perspective. This approach lends itself to the study of individuals in complex realities, and is entirely consistent with the theoretical position adopted for the study as a whole. It is seen as a way to getting to the 'insider perspective' – the reality behind the subjective experience of young students.

The process of analysis followed the approach described by Gee (2011) which involves a series of steps and iterative passes through the transcript examining and re-examining the data from multiple perspectives. This style of analysis is hermeneutic in its approach. Indeed, the researcher is trying to make sense of what the participant is saying, who in turn is trying to make sense of their world, which Gee (After Smith & Osborn) labels 'double hermeneutic'. The process acknowledges, therefore, that the researcher is an actor, and will have a perspective. So the impressions of the researcher are acknowledged as legitimate inasmuch as they arise from and attend to the participants words, rather than being imported from outside. It is called interpretive since the researcher is making a legitimate and informed attempt to try to understand the lifeworld of the individual as much as possible from the inside, whilst acknowledging that there are potential pitfalls if the interpretation strays too far from the words of the that individual. In this way, the portraits that follow will attend to, for instance, the style of language used by an individual (e.g., Helen's use of the word 'actually') or repeated expression of identity (e.g., Anna's description of herself as a 'D').

THE INCIDENCE OF DISAFFECTION – WHOLE SCHOOL COHORT STUDY

I was interested to be able to acquire more systematic quantitative data from a whole year cohort in order to understand the incidence and nature of motivational and emotional variables across a more representative population (see research questions). The opportunity arose to survey a whole year 9 cohort (n=208) at a comprehensive school. It is a comprehensive foundation 11–19 school in the UK with approximately 1300 pupils. The proportion of pupils who take free school meals is described as 'average' in the 2011 Ofsted report, and only 6% of pupils are from ethnic minorities. The school was rated as 'good and rapidly improving'. Some quantitative results of this survey will be reported here.

In just two visits to the school I was able to survey the whole of the year 9 population of this school. Qualitative responses were added to the quantitative survey. The survey was concerned with the incidence of negative affect, and was designed to provide contextual data on the landscape of disaffection from a larger scale perspective.

In their major study of pupils' attitude to mathematics, Zan and Di Martino (2007), identified three overriding aspects of student affect, and these were used in this survey. Students were asked to rate the degree to which they agreed with these three statements, on a 4-point Likert scale (the points representing 1 – 'not at all'; 2 – 'a bit'; 3 – 'sometimes'; 4 – 'a lot'):

- *I like mathematics*
- *I can do mathematics*
- *I am satisfied that I get what I want from mathematics*

Table 6. School K survey responses

	Not at all (%)	*A bit (%)*	*Sometimes (%)*	*A lot (%)*	*Mean*
Like	32 (15)	60 (29)	95 (46)	21 (10)	2.5
Can do	7 (3)	31 (15)	116 (56)	53 (26)	3.0
Satisfied	9 (4)	56 (27)	90 (43)	52 (25)	2.9

The table shows the number (and proportion) of students in each response category, for each question. How these figures are evaluated depend to some degree on the perspective. Only 10% of these pupils like maths a lot, but even that might be more than expected. 44% (29% + 15%) hardly seem to like it at all, with an additional 46% only liking it sometimes. More students appear to feel they can do mathematics than like it, with 26% reporting that they can do it 'a lot'. But that still leaves 74% who can do mathematics at best only sometimes. The 25% of pupils who are satisfied 'a lot' is encouraging, but this also leaves 75% of pupils with at least a degree of dissatisfaction.

Since the school sets groups by ability in mathematics within each half year (labelled 'K' and 'S'), we can address the question of whether, or to what degree, pupils in lower sets did (or did not) experience negative affect more than those in higher groups. This is an interesting question since it is sometimes assumed that lower attainment will lead to more disaffection, even though it is known that students in higher-attaining groups can also be disaffected with mathematics. A comparison can be made between the data from the groups in each half year. The scores below represent the percentage of pupils who reported '1' (not at all), or '2' (a bit) to the three items. This can be viewed as a blunt measure of negative affect.

Table 7. Percentage of pupils in each group reporting negative affect

	S1	*K1*	*S2*	*K2*	*S3*	*K3*	*S4*	*K4*	*S5*	*K5*
Don't like	64	24	67	29	27	60	50	43	36	50
Can't do	14	0	33	7	18	15	25	29	18	33
Not satisfied	59	6.9	33	3.6	18	40	38	48	45	58

K1 and S1 are groups at the same, high level, but they have very different scores. A higher proportion of students in S1 and S2 don't like mathematics than in any of the 'K' groups or than in groups S3, S4 or S5. Apart from groups K1 and K2, lack of efficacy (can't do) appears to be evenly spread across the ability range. Also, whilst only 14% of S1 pupils report low efficacy, 59% report dissatisfaction suggesting a non-simple relation between these two variables.

It can be seen that the most negative affect in terms of attitude ('like') are in groups S1 and S2, with groups K1, S3 and S5 having the least. In terms of efficacy ('can do'), S2 and K5 score the highest (from the perspective of negative affect). For the satisfaction scale (where low scoring suggests dissatisfaction) pupils in K1 and K2 seem to be much more satisfied than other groups. S3 and S5 seem to have less negative affect than one might expect, but S1 and S2 seem to have significantly more than one would expect.

We can conclude that all three measures appear not to decline according to level of attainment. But since parallel groups at the same attainment level can have very different scores, this suggests that it is the class or group itself that is the major determinant of pupils affective experience of mathematics. The scores seem to relate to teacher/group more than level.[1]

Qualitative Data

Whilst administering the questionnaires, within the time allotted, and after the questionnaires were completed, I had the opportunity to ask the students (who were surveyed in two half year groups) to write briefly their answers to two questions:

The most frequent or strongest emotion that you feel in mathematics classes

One sentence that sums up your feelings about mathematics

The questions were not 'leading', since the pupils were only told that I was interested in their opinions about school mathematics. Since the number of students was 208, and all students in the year responded, the results can be said to be representative of mathematics students of this age. This data is useful in gaining an understanding of how prevalent aspects of disaffection with mathematics are within the population of that age.

For the single emotion-word response data, care had to be taken in organising and analysing the data. For instance, board, bord, bored, boredom and boring were all taken to refer to the single emotion of boredom. Multiple variations on other terms were also similarly consolidated. The words were then classified in a simple 'positive', 'neutral' or 'negative' manner. Although this is a fairly simplistic way to organise the data, it does have meaning within the context of this study. The results are shown below:

Table 8. Emotion word responses

Positive	*Neutral*	*Negative*
37	29	135

This is a dispiriting result, and even more so since the cohort includes the full range of ability. It suggests, at the very least, that mathematics is not a pleasant experience for many students, for much of the time. Nonetheless it is also important to point out that it is not necessarily the case that pupils who report boredom are disaffected. To be strongly disaffected a pupil would have to report experiencing a whole range of negative or adverse affective responses.

Individual results include:

Anger 11, Boredom 68, Confusion 10, Stressful 8, Depressing 5

On the other hand, 'Happy' was chosen 21 times, but 'Fun' only once.

The 'boredom' score here is consistent with boredom being the highest scoring negative emotion on the TESI-ME (see below), although the population there is very different. Such results confirm data presented in the literature review on the incidence of negative affect in the school population as a whole.

In the two top classes (labelled K1 and S1) 30 pupils (17 + 13 respectively) out of 51, which is well over half of pupils, reported negative emotions, of which 19 were 'bored', whilst 20 pupils (5 + 15) reported positive or neutral emotions. It is worth noting the very different numbers of pupils in the two classes reporting positive emotions, suggesting again that classroom climate is an important factor influencing students' affective experience of mathematics.

In the two bottom classes (labelled S5 and K5) 13 pupils (2 + 11) out of 23 reported negative emotions (about half) of which only 7 were 'bored', whilst 9 pupils (8 + 1) reported positive or neutral emotions. Note again the very different proportions of pupils choosing positive or negative emotions in the two classes.

On the basis of this data alone, it can be reasonably concluded that the group itself is more important as a factor in determining aspects of pupils affect than is the level of ability of the pupils in it. This is an important finding. Some caution needs to be applied in generalising from this conclusion, however, due to the simplicity of the data, and the small numbers in each group. On the one hand, the cohort represents the full range of ability. On the other hand since only 208 pupils were surveyed, no attempt is made to underwrite the statistical significance of the results. In addition, a single one-word response does not represent a full examination of these pupils affect in relation to school mathematics.

The descriptive passages were also analysed, using the framework developed by Zan and Di Martino referred to above. It is of interest in the current study to look at the data from the point of view of positivity or negativity. Thus the short passages were examined and classified as 'like', 'don't like', 'can do', 'can't do'. An example of a 'like' statement would be 'maths is exciting and you can learn a lot from it.' A typical 'can do' statement would be 'I am good at it.' A typical 'don't like' statement is 'I don't like it and don't enjoy doing it.'

Statements about the nature of mathematics were captured separately, but also classified as either positive or negative. Although simple, this approach yields instructive results.

Table 9. Statements classified according to theme (Percentage of pupils in each group)

	S1	*K1*	*S2*	*K2*	*S3*	*K3*	*S4*	*K4*	*S5*	*K5*
Like	5	17	26	32	18	10	19	10	9	17
Don't like	32	14	22	14	5	25	19	29	18	8
Can do	36	24	15	11	55	25	25	33	55	25
Can't do	14	14	37	7	9	15	19	29	9	0

There are some difficulties and ambiguities in this process. Null returns were ignored. Some statements could not be classified within the framework since they were neutral. The most difficult ones to deal with were those that could be interpreted in more than one category. These included those of the form "I like maths when it is easy, and I hate it when it is hard." In these cases it was classified as both 'like' and 'not like'.

The individual numbers in each cell of this matrix are too small to offer categorical interpretations. However, some points do stand out. As with the single emotion words, the numbers in each of the variable categories differ widely by group. So, for instance, there are many more pupils that like mathematics in S2 and K2 (26% and 32% respectively) than in any other group. Similarly, there are more pupils that report 'can do' mathematics in group S3 (55%) and S5 (55%) than in other groups. Only 4 groups (S1, K3, K4 and S5) have more pupils that 'don't like' than 'like' mathematics.

In all groups apart from S2 (including each of the four lower groups S4, K4, S5, K5) there are more 'can do' than 'can't do' reports. Classes differ in the relative proportions who 'like' and 'don't like' mathematics (and in terms of which one predominates in that group). The evidence doesn't support the assumption that higher or lower groups like or don't like mathematics more than the other. Positive or negative affect (liking, not liking) and competence (can or cannot do) do not appear to be related to the level at which one is achieving. Put another way, students in higher groups appear to be as likely to not like, or feel they cannot do mathematics as students in lower groups.

The Nature of Mathematics

In most cases it was quite easy to identify those statements that related to the nature of mathematics rather than to affect or competence. These statements were split evenly between positive and negative. In terms of positive statements, the most common were about the general utility or value of the competence:

> *It helps in life in some situations*

> *I think maths is life changing and it can help you in the future*

A subset of these related directly to the exchange value of a good qualification in mathematics:

> *It's an important subject and you need a good grade to succeed in further education*

There were also some comments about mathematics being of value in its own right:

> *Maths is good for making you think*

> *Maths is a tool we can use to solve problems*

And more intriguingly:

> *Maths isn't very useful later in life but it challenges me which is a good thing*

The negative statements include those that reflect the nature of mathematics as experienced by them. These include descriptions like 'hard', 'complicated', 'confusing', 'lists of tedious questions'.

Other negative statements related to the perceived lack of importance or utility:

80% of the time completely useless for my future (presumably said without irony!)

Sometimes I think it's pointless

It isn't the primary purpose of this study to investigate in depth the epistemological beliefs about mathematics held by pupils, and no claim is made that this data represents a comprehensive examination in that way. However, pupils' views also influence their affective landscape, and it is interesting to have some idea of this broader picture, as exists in year 9, to compare and contextualise the views that will now be examined in more depth of the subset of pupils who 'fail' mathematics at GCSE, and need to study it further at College.

INCIDENCE OF DISAFFECTION – QUANTITATIVE DATA FROM THE MAIN STUDY

The TESI-ME

In all, 184 returns were acquired. These comprised of 74 from College R, comprising 5 classes, 51 from College G in 2011 from 4 classes, and 59 from College G from a further 5 classes from the following year. 99 of the respondents were female and 85 male.

All scores were rated on a Likert scale of 1 to 7, where 7 represents 'very much'. The mean score for stress was 4.31, and 4.66 for effort.

Since scores of 5, 6 or 7 can be seen to represent significant negative emotion such that the student can be considered as at risk of disaffection, number and proportion of students reporting such scores are shown below:

Table 10. Number and proportion of students who scored 5, 6 or 7

Anxiety (%)	*Boredom (%)*	*Anger (%)*	*Sulking (%)*	*Humiliation (%)*	*Shame (%)*	*Resentment (%)*	*Guilt (%)*
35 (19)	73 (40)	49 (27)	37 (20)	25 (14)	28 (15)	27 (15)	25 (14)

The high scores for boredom, as has already been pointed out, were to be expected. However, the prevalence of other negative emotions has not been well accounted for in much literature on affect in mathematics education. More students reported high scores for anger than for anxiety.

The eight scores for each student were totalled, and these can be considered to be a crude index of individual negative emotion concerning mathematics. The ranges were 8 to 53 (out of a possible maximum of 64), with the mean at 21.9. 6.5% of students scored a total of 40 or more, and 17.4% scored a total of 32 or more (representing an individual average of 4 or more per emotion). It can be concluded that such high scores show evidence of a distressing emotional experience of mathematics for these students. To score 40 or more suggests that mathematics is an unpleasant experience for the students in that category, associated as it is with consistently high scores on these negative emotions. No attempt is made here to generalise these claims to the whole population of students in this age range. However, it is worth bearing in mind that they may be somewhat representative of the population of students who have failed to achieve grade A to C in GCSE mathematics, and that population itself represents a large number of people.

There is an interesting pattern in the data that should be remarked upon. Firstly, the mean scores for somatic emotions (anxiety, boredom, anger and sulking) are higher than those for the transactional emotions (humiliation, shame, resentment and guilt). 42 students (23%) scored 1,1,1,1 for these four transactional emotions Only 4 students scored 1,1,1,1 for the somatic states. It is not easy to find an interpretation for this pattern. It could, of course, just be that these negative transactional emotions are not aroused as much in the context of mathematics classrooms. It may also be that for many students, the transactional states are less easy to recognise or compute. This in turn could also be that, because they relate to our experience of the external world, and in particular, social relationships, that these emotions take time and experience to develop fully in mature adults. It is possible that in young, emotionally and socially naïve students, that they have not yet fully developed. This might be an interesting issue for further research.

The validity of the instrument derives from the associated constructs derived from theory. However, reliability, and other psychometric properties of the instrument, are not assured in this context, and so generalisations based on such are not made. Rather, the instrument is used descriptively, and is seen to have value in terms of the context of the interview, as described above. Further, the use of the instrument cannot be said to have confirmed that the emotions as used in the instrument relate to states as predicted by Reversal Theory, and this is particularly true of the transactional emotions, which the evidence here suggests are difficult to study with the population of interest here.

The Card Sort

In all, 48 interviews were conducted, including 33 with College students and 15 with school pupils. In every interview, students/pupils were invited to choose cards from both the green and blue sets, that represented something relevant in their own experience. Perhaps the first surprise is that this was not at all difficult to do, since evidence of positive experiences is rarely ascribed in research to disaffected students. In all of the interviews, only two people failed to choose a card in one category, and no-one was unable to choose any cards at all. Most chose a couple of cards in each category.

The significance of this is that it establishes straight away that even students who are highly disaffected have these islands and episodes of positive experience in their relationship to school mathematics. What the cards enable us to do is to investigate qualitatively what these positive experiences are. This will be reported primarily within the context of the case studies and the cross-case qualitative analysis (below). However, it may be instructive to provide a count of the choices of the students in the interviews. Tables 11 and 12 give a quantitative picture of those choices by card and by institution.

Table 11. Blue card choices (Emotion words)

	College R	*College G*	*School G*	*total*
Grateful	1	5	3	9
Modest	1	1	2	4
Proud	4	14	9	23
Excitement/curiosity	2	9	6	17
Relaxed	3	16	5	24
Mischievous/playful/naughty	3	7	4	14
virtuous			1	1

Table 12. Green card choices (Motivational values)

	College R	*College G*	*School G*	*Total*
Sense of purpose/Importance	1	2	1	4
Freedom	2	9	3	14
Powerful/in control/ competitive	2	3	1	6
Sense of duty	1	5	4	10
Sense of achievement	2	14	8	24
Fun/enjoyment	3	6	5	14
Part of the group or club		9	5	14
Helping others	3	18	7	28
Feeling cared for		3	3	6

Students at each of the institutions made as many positive choices in proportion to the number of interviewees, although Students at College G made more positive choices compared to college R. In qualitative terms, nobody in college R reported feeling cared for or feeling part of the group, and this is in contrast to both school G and College G. No statistical significance can be claimed for this result, and so interpretation should be guarded, but the result would fit with my own assessment of the classroom climate in the respective colleges.

In terms of the blue cards (which represent emotions), 'relaxed' and 'proud' are the most popular choices, being selected by 24 and 23 students respectively (representing nearly half of all students interviewed). This is followed by 'excitement/curiosity', chosen by 17 students. These figures emphasise the tangible motivational and emotional benefit to students of being able to successfully complete classroom tasks. It may seem self-evident, but ensuring that success is a regular part of students' experience of mathematics is a vital motivational tool.

The most chosen green card (and in the whole card exercise) was 'helping others' (chosen by 28 or over half of all students). This result is significant in that the notion is hardly reported in the research literature in mathematics education. It appears to be as significant for school pupils as for College students, and this leads to the suspicion that it is a quite general motivational factor in the learning of mathematics. Since it is not widely reported, it is possible that teachers don't prioritise it as an important component of the motivational climate in their classrooms. Indeed, in this study, students often characterise their experience of school classrooms as being solitary, individual and silent activity, as reported in the TIRED framework of Nardi and Steward. 'Feeling part of the group' was also chosen by 14 students, and this, taken together with the results for 'helping others', provides strong evidence of the importance of social activity in the learning of mathematics.

The second most chosen green card was 'sense of achievement' (chosen by 24 students or approximately half). This reinforces the point made above about the importance of success as a motivational factor in mathematics classrooms. Looking across the cards, 'excitement/curiosity' (scored 17) together with 'fun/enjoyment' (scored 14) and 'mischievous/playful/naughty' (scored 14) all relate to different aspects of the playful motivational state, and they appear consistently and strongly in students narratives about motivational and learning mathematics.

One other result of note is that 'freedom' was chosen by 14 students. In the context of mathematics classrooms, this is usually interpreted as being able to do something in the student's own way, or with a method originated by themselves, and different to the method shown by the teacher. That it is motivationally satisfying to do this is a strong argument for constructivist or student-centred pedagogy.

Finally, it should be noted that these results should be held in perspective. They represent the motivational aspects of positive experiences of school mathematics. In this sense, they are not a rating of the importance (or perceived importance) of the notion. Qualitative evidence presented subsequently will show that other aspects of motivation and emotion are also important. To take an example, 'feeling cared for' was chosen by just 6 students, but this does not mean it is not important, rather that it is not being experienced (in the positive sense) by many of these students.

The value of the card sort exercise, and just like the TESI-ME, is that it is a guided stimulus that creates data that is an occasion to discuss motivationally and emotionally important ideas with individual students. However, taken as a whole the data also provides a quantitative picture of aspects of motivation and emotion that are relevant in mathematics classrooms, and are at the same time theoretically grounded. In this way it can be claimed that they widen and deepen the language and the construct space to enable these matters to be investigated and discussed.

NOTE

[1] This theme will recur later, and will be discussed further. It has only very recently come to my attention that Noyes (2012), in a paper entitled 'It matters which class you are in: student-centred teaching and the enjoyment of learning mathematics', gives a quantitative affirmation of this result.

CHAPTER 6

THE CASE OF ANNA

IMAGES OF DISAFFECTION

There are a number of ways that the data in this study can be analysed and presented. Since the primary aim of the study is to understand better the phenomenon of disaffection from the point of view of the subjective experience of students, the first section of the results will be reported as a set of case studies. In this way, the individuality of the complex and dynamic nature of the relationship that these students have of their experience of learning (or not learning) mathematics can be fully represented.

Each of the case studies follows broadly a similar pattern, in that some background is offered to their early experiences of mathematics, followed by an account of their decline into disaffection. The experience of disaffection is then described, which is then balanced by evidence of their more positive experiences (often elicited by the card sort). Reversal Theory is used to provide explanatory insights, either within the context of the story, or (for instance in Anna's case) in a separate section since this aids interpretation of the data. In places, narrative quotes from the students are used as side titles to personalise the structuring of the account. Variations to this broad format are used, when the story seems to justify it to reflect the particular emphases and nuances of the individual story.

These cases were selected since they represent unique but very different, and highly personal ways of experiences aspects of disaffection. It is not claimed that they represent all possible aspects of disaffection, or that they represent examples of any typology of disaffection. In a sense, any or all of the interviews could have been reported as individual cases, but space would not permit this. The criterion for selection, then, was pragmatic, based on individuality and vividness of the stories. I am aware that I could have chosen other stories, and that other researchers may well have done so with the same data. The contention here is that each represents a valid representation of disaffection, of interest to the topic at hand.

WHY AREN'T I AN A OR B ANYMORE: THE CASE OF ANNA[1]

Anna is a complex and interesting character. That she is disaffected with mathematics is shown by her life history. It starts positive but low, and even receives a boost in year 8/9, but thereafter drops dramatically into the negative. Having got on with mathematics unremarkably in her early school career, and made progress, events had a dramatic effect, and despite her competence, she 'failed' (in her own terms) at GCSE (national examinations at aged 15 years), and this failure colours her whole subsequent experience of mathematics.

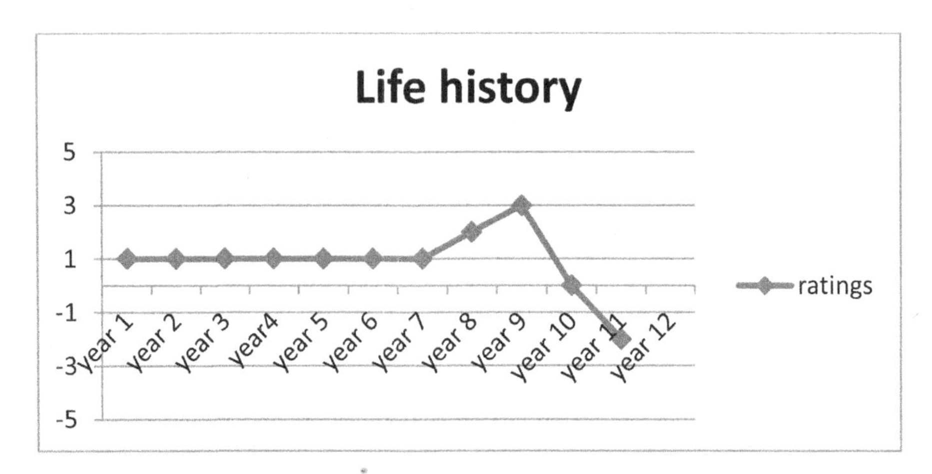

Figure 6. Anna's life history

She presented at interview as bright, very thoughtful and full of curiosity. She was very small, and very gentle, but lively, and her manner belied her strong views. She was articulate about her disaffection and this made her a particularly good subject.

Early Years – A Time of Innocence

By her own account, Anna got on with mathematics in the early years in a quite unproblematic way. She didn't regard mathematics as in any way special, or different or even difficult. Indeed, she seems almost surprised herself that she hardly recognised it as a separate subject.

> it was just something …I just thought we had to do…like at school…when you're at primary school you have to do geography and everything…you don't actually think about it…it's just work…you just do it.

She gets on with it, and she finds it 'not too hard.' We see here a quite conforming attitude and approach (although this was to change). At this stage, her given epistemology of mathematics is quite limited (and this was to change also). However, she was able to do it, even though she didn't see it as her favourite subject. She doesn't particularly enjoy it. In retrospect, although she doesn't offer this as a cause-effect explanation, this is a time of innocence where, in her words:

> At primary we just get stuff to do…not too hard really…we don't get like sets or grades like school…like in college you get grades like 4b or 4c or whatever…like A's and B's and stuff like that. We didn't get grades then…so it was just like doing work…

Not getting grades makes it just like doing work: This is an interesting juxtaposition. 'Just' doing work sounds so unthreatening. But it also suggests a fairly malign view of the effect of getting grades. This is significant because it is precisely these 'grades' that come to dominate her experience later on, and represent a major cause of her disaffection.

She obviously does well, reporting that she did well in her SATS (national tests) and was promoted to the second set. And although maths didn't excite her interest too much, she does report that 'shapes' were her favourite thing.

Decline into Disaffection – Innocence Lost

A key watershed in her experience of maths is that, having been told that she was an 'A or B student' in year 9, she then suddenly found herself demoted to the status of 'C/D'. This is a seminal experience, and one from which she still has not fully recovered. As she says, "after that, it just went down and down." Maths had never been one of her priorities, but she had still thrived and succeeded. It all changed quite suddenly and unexpectedly. She describes that she had trouble at home, and then she did poorly in an exam. In a very short course of time the world came to view her in a different light, and this leads her to view herself in a different light also.

One of the interesting features of this account is that she shows a strong sense of the relationship between her behaviour and the outcome. She didn't practice; she didn't do as much homework – she is taking responsibility for the outcome. This is an important part of her style, and will be commented on again later. Here she gives part of the explanation of the decline:

> I just lost interest (pause...) in it...I dunno...I was never like interested in it... (pause)...I dunno...one of them things...practice...you have to actually like it.

The sense of loss at being re-classified seems to be a mystery to Anna, and one that clearly hurts her deeply.

> cos if in year 9...I'm told I'm going to get a B...and in year 10 I'm now told I get a C/D...that's not good is it?...

Anna does try to recover both the grade, and her position and status in maths, but she has a hard journey ahead. She failed GCSE. She now says she is just 'doing it for the grade', but there are hints that there is more to it than that.

The Experience of Disaffection – 'Why Aren't I an A or B Anymore'?

It has already been noted that Anna has a range of negative feelings about maths due to her experiences at school. But these carry over and colour her experience of college. On the TESI-ME she scores 6 for stress. She explains:

> yes...sometimes I just don't get it...and don't get why we have to do it

More specifically, she reports medium-to – high anxiety (score 5):

> apparently I've got anxiety my doctor says…I'm not too sure what it is…but panic…ignoring that…I do sometimes panic...when you get...exams…kind of freak out like shit…I don't think I'm gonna pass this…

When asked about exams and stress, she articulates the fear very clearly. Again, she feels intimidated by the labelling that grades represent, and this very public comparison to others is also damaging. But part of the meaning of that for her is that it also tells her she hasn't done the best she can. Again, she takes ownership of the outcome – it's nobody else's fault but her own. It appears she feels guilty for letting herself down, and as a result, she feels 'kinda dumb', or more poignantly, 'I feel like a really bad D.' Again, she returns to the comparison that she was told she was a B, and she once felt like a B, and now she is a D, and she can't understand why. No wonder she feels anger and resentment at this situation. She also feels guilty that she let her domestic situation affect her so much. Even this guilt is compounded by her adverse comparisons to others – in this case her boyfriend, who also had ('even worse') problems than her, but still did well in his maths. In terms of her poor performance then, her first instinct is not to blame others. She shows humility in her questioning of herself:

> not being able to get that B…that's what makes me angry…and I don't get why I can't get it…is it the teachers…is it me…I dunno…I understand the work…I just can't seem…

Anna's version of anger contrasts somewhat with descriptions in this data from other students. On the whole, other students describe anger as an in-the-moment phenomenon, that is triggered as part of a negative sequence of events and their associated motivations and emotions. Since it is uncomfortable, actions are usually taken to mitigate the anger. The anger that Anna describes is more prevailing since it endures over time. The emotion is stabilising into a negative orientation to the world: a crystallised way of interpreting certain phenomena. It seems that she does not find maths at college very easy and so now she has to work at it.

Anna from a Reversal Theory Point of View

To understand a little more about Anna's experience of disaffection it is useful to examine her motivations and emotions from a Reversal Theory point of view. Part of the depth of her disaffection arises not just because of the intensity of the feelings, but also because there is a gap between her motivational needs and the reality in relation to many of the motivational states. This means that, almost whatever state she is in, she is dissatisfied – she is trapped into negative evaluations of her situation.

When examined from a Reversal Theory point of view, perhaps one of the most interesting aspects of Anna's motivational landscape in relation to her disaffection is her transformation from *conformity to rebelliousness.* It has been noted (above) that she was fairly conforming in relation to school mathematics in her early years. The overall impression from her description of primary school and early secondary is that she did what she was 'supposed' to do. As she put it: "You just get work to do and you do it." There is no hint here that she questioned her role or obligations. This changes significantly. In her account of her experience from year 9/10 we see a very strong expression of rebelliousness in Anna. This rebelliousness is clear in her descriptions of her anger and resentment at her situation. This anger arises in the serious-rebellious state combination, when arousal is high. But it is also clear in her general questioning demeanour. This rebelliousness manifests in a number of ways. It manifests in conjunction with the serious state in her perception that she was a B, and is now a D. Her sense of unfairness at this is expressed in her anger (as quoted above). That is *not* supposed to happen. Some rule or expectation, for her, has been broken, and this excites the rebellious state. One can hear her rebelliousness in her off-the-cuff critique of how the notion of research is presented in her fashion classes, as well as her evaluation of other aspects of her experience of maths teaching. Her fight to get put into the booster classes is a clear expression of disapproval and rebelliousness. It is also interesting to see how often she displays her more mischievous side, which can be interpreted as the playful and rebellious state combination (paratelic negativism). This is demonstrated best in her turning the tables to interview me, but pervades her attitudes and approach, and is likely to be a more socially acceptable way to express her negativism than anger.

It is also instructive to look at the notion of *purpose or achievement*, which is the driving value of the serious motivational state. For Anna this operates in both the foreground and in the (longer term) background. In the context of College mathematics, the serious motivational state relates to the successful completion of tasks. The evidence here shows that she takes this seriously (in the everyday sense of the word). She wants to succeed, and she applies effort (but not particularly interest). Her approach can be described as serious-mastery, since she not only wants to complete the tasks, but she has a competitive mastery orientation that means she needs to do better than those around her. So she works alone to achieve this, and indeed, she is able to derive some sense of satisfaction in this mode. She reports relaxation (the pleasant emotion associated with satisfaction in the serious-conforming state combination) when she has done something well. She reports a sense of pride (the pleasant emotion associated with 'winning' in the self-mastery state combination) when she gets good scores on tests, and she describes her pleasure at being able to help others who don't 'get it'.

But this in-the-moment foreground activity is balanced by the longer term background need for achievement which is based on a competitive interpretation of success. For some students, satisfaction in the serious state can be gained by an appreciation of the importance or utility of school mathematics. But in this sense, Anna is ambivalent, although her sense of this importance, and indeed, her notion of what mathematics is, develops in sophistication during her college years. It has already been noted that, in her primary years, she classified maths as 'just work', and she was hardly able to differentiate it from other subjects.

> I couldn't actually tell I was doing maths in primary school…you understand?
>
> This unconsidered epistemology of mathematics may in part explain why she 'sleepwalked' into a situation that she later could not control. We have already noted that she shows little sense of enjoyment of maths (apart from 'liking shapes') in her early years. The way she talks about maths currently suggests that, along with her developing maturity through her life experiences, and experience of other subjects, there is an emerging sense that there is somehow more to mathematics than she has previously given credit for. Statements like 'Maths…it's got so many things to it', and 'it's like art', give a hint of this. It is stated more explicitly in:
>
> Maths is like a pivotal subject isn't it? It's like physics and stuff are based around maths aren't they?

Much of this developing sense of the importance and scope of maths comes from her reflections on her own experience:

> cos I wanna be a fashion designer…I make the paper block…which is basically like the net of a cube..but the dress version…does that make sense?... and I have to add seam allowance…which is the flappy bits you add to allow for glue…stuff like that…so until I actually did that…I didn't think I'd need maths…does that make sense?

In these ways her experience of the serious state in relation to school mathematics develops over time. It starts in primary as being 'just work', but unproblematically so since she could do the work. During and after the difficult years, the serious-conforming state has become broken for her, tainted as it is by the sense of failure. Her experience is also massively coloured by the humiliation of the damaging effect of her feeling labelled. She talks eloquently of her sense of purposelessness – of not being able to envisage a successful outcome in this frame, or even of having, at times, a sense of 'why bother' with this. She clearly has an overriding need to achieve a pass at GCSE, but this is a frustration, since it has so far escaped her. She carries this need going forward, but suffers anxiety at exams (an emotional consequence of the serious-conforming state combination together with high arousal), and feels she is likely to fail. However, during her college years, she begins to gain an appreciation (and potential satisfaction) from a more positive interpretation of the use of mathematics.

On the other hand, she displays a good deal of evidence of being regularly in the *playful state*, but not in relation to her activities in the maths class, where she seems unable to find expression for it. She rarely expresses any paratelic enjoyment of work in maths, except perhaps when describing her pleasure at discovering that nets are useful in fashion design (an interest in 'shapes' that she carries from her primary school experience). Her curiosity, her interest in my past and motives, are all evidence of her playfulness.

In terms of the transactional emotions, her experience is driven by the motivational state of *mastery*. In her case, unfortunately, this is often experienced as mastery-losing. When she is in the self-mastery (losing) state combination, this is experienced as humiliation, and there is plenty of evidence of this. When it is experienced as self-sympathy (losing) it is experienced as resentment (i.e., I am not getting what is due to me – success). Her sense of mastery is sharpened by the fact that she uses external measures to evaluate her mastery position. Where she stands in relation to other people is very important to her. She is in fact, very competitive. She describes classroom strategies that increase her chances of outperforming the other classmates.

> so… I feel quite competitive…like I want to do better than everyone in this class…but I don't know if I will. (LAUGHS)

Even within the context of her positive experiences of *helping others* (other oriented mastery), she recounts:

> it makes you feel good…because you get it and someone else doesn't…so you feel a bit proud…but if you're helping them…you feel a bit nice about it… does that make sense?

Notice the conflicted response here. Beating someone else makes her feel proud. Does that in turn make her feel 'not nice'? Her 'does that make sense' comment hints at her ambivalence about feeling good because someone else doesn't understand. This ambivalence carries over into the notion of helping other people, which clearly gives her satisfaction. But she is at the same time aware of her pleasure at seeing others do not as well as her. Here is her description, illustrating this ambivalence, and the reality of the pull of different and contradictory motivational states:

> It's not like I can do it and they can't…it's like I can (my emphasis) do it… and they can't…it's not like they're bad…not like ohmigod I got better than her…I don't really care what they got…I got better than them…it's not about them…it's just a good feeling isn't it?

For her, this helps to explain why helping others is not only a good thing to do, but also useful to her:

> because apparently if you teach someone you understand it more yourself… cos you're repeating it…cos maths is about practice…you're technically just repeating it…it does help…if you help other people you learn yourself…

From the evidence here, the progression appears to be:

- Self-mastery – competitiveness, and an attempt to do better than the others, leading to a sense of pride (if successful);

- She then notices the lack of progress in others (other-sympathy), which may trigger a sense of guilt;
- This causes her to help her classmates (other-mastery), which in turn results in her feeling 'nice' (self-sympathy).

In this way, helping others can be seen to be a motivationally rich activity for Anna, since it satisfies a range of motivational needs – even though some of them are contradictory.

A Chance to Right the Wrong

Given the analysis above, it is now possible to evaluate Anna's motivational relationship to mathematics in her current situation. Although Anna insists that she still doesn't like maths, it seems that passing is the only way she can recover the sense of achievement and competitive mastery that she lost in year 9 and 10. As she puts it – 'I'm going for the grade'. She is at pains to emphasise that she does not need the maths as she could get a job without it.

But making progress, being able to do the tasks, does provide her with sense of achievement and positive emotion and satisfaction. Being able to 'do it' and getting good marks makes her feel proud – particularly when others can't do it. It appears to restore her sense of competitive self-mastery.

> when I get something right…I do feel quite relaxed…like when you go shopping or something and you can actually work it out before you go to the till…you're quite grateful you learnt it…does that make sense?

She is willing to put effort and determination into the search, as shown in her efforts in 'begging' the teachers to let her do the booster classes. Her description of her sitting alone, so as not to be distracted by others also shows her competitive determination to do well this time around.

She also has a sense of realism. When asked what she would do if she needed A level maths (a specialist, post-compulsory qualification) to get a great job in fashion, she replies:

> eh…but I would do it…I would probably put everything I had into maths… you can…it's just time and effort…I'm not too sure…it's just a theory but… it's like business isn't it…business is time and effort and that's how you get a lot of money.

In fact, she use s this notion of practice to explain to herself how she fell behind in the first place:

> I kinda lost a lot of practice as well…cos things weren't great…so…obviously I wasn't doing as much homework as I should…

This notion of practice, together with her strong sense of personal responsibility leads her to change her mind about how success is achieved:

> I used to think that people are just born smart…but then I kinda got my head round it and now I just think…(HESITATES) ..it's time and effort…you're not just born smart…it's time and effort… it's how much you put into it… cos anyone can get good grades…you can…like if you went home and if you went through everything everyday…and you were that organised…you can get good grades…I think it's time and effort.

Anna's story contains a dominant narrative, and this is one of her anger and anguish at being classified as a 'C/D' and her corresponding performance at that 'failed' level. This experience is in stark contrast to her early, innocent years, before labelling and failure, when she was able to get on with school mathematics in a seemingly untroubled way. However, the consequences of her failure go deep, and provide the motivational and emotional context in which all of her subsequent experiences in filtered and interpreted.

NOTES

1. All of the names have been altered to protect the anonymity of the students.
2. This chart, and those for the other cases, have been recreated as charts for convenience, as the originals are very cumbersome to scan and illustrate.

THE CASE OF HELEN

Helen scored quite highly for overall stress on the TESI-ME (4 out of 7) with high scores for boredom (7), anger, humiliation and shame (6). These scores suggest a high degree of negative affect even compared to some of her peers in this sample. Her life history scores are medium with increases in year 5 and 7 (at aged 10 and 12 years), and then a significant decline in year 10 (aged 15 years). Helen's story can be told in relation to four key themes.

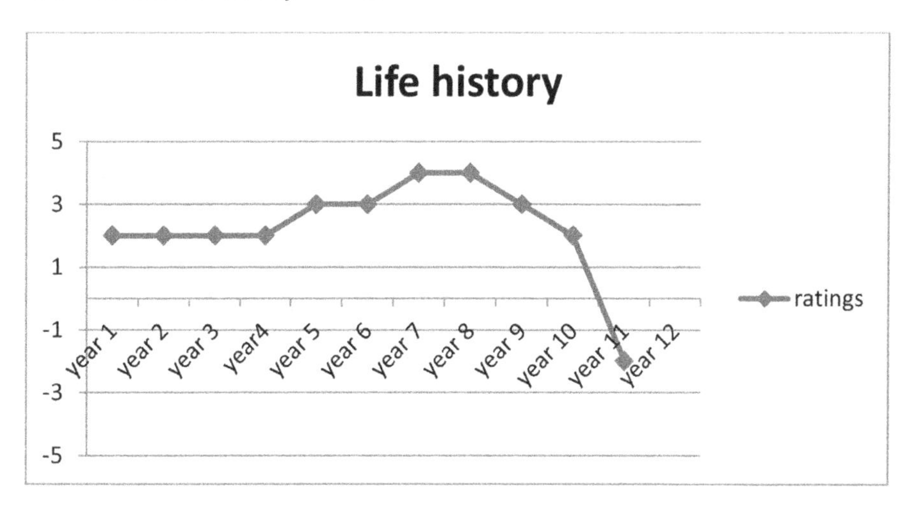

Figure 7. Helen's life history

Decline into Disaffection

Although Helen reports a positive relationship to mathematics over many years, her attitude to the subject is ambivalent throughout as is evidenced by the tentative way she uses language ("It wasn't that bad..."). Her decline into serious disaffection starts in year 9 when the maths gets harder and (as she sees it) the teaching is poor.

> mmm...basically we'd just go over the same stuff and we'd never learn...like he never explained it...and he left half way thro' the year so we always got supply teachers.

Her description of the confusion and procession of supply teachers being constantly swopped around throughout year 10 is heart-rending. As she says:

> we had no one to build us up…and then I didn't get entered into the year ten exam 'cos everything got messed around.'

By this time she was lost. "I just hated it." and "I'd just switched off by then."

One of Helen's key narrative themes is confidence. Her relationship with mathematics ebbs and flows in tandem with her confidence. And as well as being volatile, there is no sense of agency or internal regulation for her confidence or competence – it won't sustain on its own without the external push from the teacher. This is evident in many examples of the forms of language like:

> ok …(pause)…mmm I know I got more confidence in year 5…'cos all my reports in year 5 say I was really good in maths

It seems that as much as teachers can confer confidence, they can also take it away.

> we had no one to build us up…

The Experience of Disaffection

In this section we look at Helen's description of the motivational and emotional landscape of the disaffection itself. In her early description she says "I hated it…I lost confidence." Later on she elaborates some of the detail. When asked about her high score on anger she provides a vivid description of how it comes about:

> It's like…I can't explain it… like nervous…like I can't do it…like you know when you can't do it…on the edge…'cos everyone else can do it…you're looking around…everyone's doing it…and I'm sitting there…

Then on the anger itself:

> Well…like sometimes when we're doing like decimals…I can't do 'em… nothing…and actually anger…and even how much she explains it to me I can't do 'em because it just doesn't register…then I get frustrated and then I'm not doing the work…because obviously I can't do it. (laughs)

In motivational terms this is an interesting sequence. We can infer from her description that, when faced with a difficult task, she is in a serious-conforming state – i.e., wanting to complete the task as set. The nervousness or anxiety she describes is the emotion that arises in this state combination when arousal is high, and it is high because of her realisation that she can't do the task. Her description of 'being on the edge', and looking at others who can do the task suggest that she feels under threat. Her self-comparison to others suggests also a feeling of mastery (losing), which in turn feels unfair and this trips her into anger (associated with a reversal to the rebellious or negativistic state). Helen's behavioural response to the anger is:" I can't do it so I don't do it…I'll sit there drawing." Later on she describes separately the sense of humiliation, which is self-mastery (losing):

> Like…humiliation sometimes…when you sit there like I say…you look back…and you think everyone else…and I can't do this and I should be able to do it.

Later on she amplifies the description:

> I was just staring at it…I don't know none of this…and I used to sit there and look at it and think…that's jibber jabber to me…like I don't know nothing…

This sense of helplessness and hopelessness when faced with a task that she can't fulfil, and the motivational, emotional and behavioural sequence that follows is a recurrent pattern for Helen, and an influential part of her experiences with mathematics. One of the problems for Helen is that it takes time and repetition for her to achieve competence. Listening to explanations often doesn't provide learning for her.

> I don't know what it is…she explains it to me…sometimes I get it straight away…she keeps saying it to me…but I don't seem to register a lot…it don't seem…

Helen also seems sensitive to the nature of the delivery by the teacher. Here she describes how boredom comes about:

> When you're sitting there and someone's got such a dull voice and you're listening to them over and over again…it's just…like…you switch off… because…erm…you don't feel like it's something to listen to…does that make sense?

I Can't Believe I'm any Good

This theme gives a brief account of the more positive experiences and satisfactions in Helen's relationship with mathematics. We have already noted the important and periodic boosts to her confidence, provided by transfer to a new school (and a new teacher) and transfer to a higher set.

> yes…because I actually went into a higher class…and obviously that gives you a boost

But we have also noted the tentative nature of that confidence. So, when describing positive aspects of her experience of maths, her language also reflects the tentative hold she has on competence. Thus we have: "I *actually* (my emphasis) went into a higher class."; "I actually learnt it."; "I actually thought I was a bit smart ...to be in one of those top classes."

Notice the surprise implied by the 'actually', that she was only a 'bit' smart, and 'those top classes' also implies a non-personalisation – not my class, but their class. Even when she is temporarily successful, she appears to feel like mathematics is a club to which she does not fully belong. She expresses this in the third person, through the experience of her mother, when she says "It's become immune to her." The notion of immunity is an interesting metaphor (as is the fact that it is immune to her, and not the other way round). It suggests the hostile nature of maths to her, and the fact that she needs protecting in order to engage with it. The evidence in the current study suggests that this is a common phenomenon amongst these students.

Despite that, she does report some positive experiences. Helen chose the cards *feeling part of the group* and also *helping others*. Both of these involve a social aspect, and in Helen's mind, both involve the notion of help: "Yeh…I like doing groups 'cos I like everyone supporting each other more." And further on:

> 'cos…obviously you know if everyone's working in that group and if everyone can't do it…so if you're working in a small group it works out better because everyone can help each other… that's why I like helping others.

Helping others also helps her to learn:

> Yeh…because when you're helping 'em…you're saying what you've just been learning…so it's helping because you're re-saying it like over and over again.

The net result is "in the end I actually learnt it." – but again, note the use of actually, as if she is surprised by this. Help can also come from unlikely sources, such as being allowed to use a calculator (even though it feels a bit like cheating):

> But we're allowed to use a calculator…so I feel like I've got more help if that makes sense… so it feels like it's a backup to help.

Other positive aspects of her experience include having fun, maths that is 'real', and a teacher who is energetic and enthusiastic. She clearly responds to an element of intimacy in her relationships with teachers:

> mmm…there's one teacher at my secondary school…I liked him quite a lot… and…in some ways he always used to boost my confidence…
>
> You get on with them (teachers) when they teach you more.

The Utilitarian Contract – "I Don't Need You... Well, I Do Need You"

Helen's story has shown that her fragile confidence and hold on mathematics was broken by her adverse experiences in year 9 and 10 at school, which led to her failure in national examinations. At the point of interview she was doing a course of mathematics that she was required to do. It would be easy to think that being a conscript rather than a volunteer would negatively affect her attitude and motivation, but the reality is more complex than that. Her ambivalent attitude, that she would like to say "goodbye for ever", but knows she can't, is reflected in the quote in the section title. In this she has been influenced by the experience and advice of her mother, who also struggled with maths to qualify as a nurse, but who stuck at it to achieve competence and success. So, despite the odds, she retains a sense of purpose and a sense of the importance of maths in her future life. When asked what she would do if she had to learn some (for her) complex maths in order to fulfil a future job role and gain promotion, she says:

> I would try it...I *would* (my emphasis) try it...I would take it...if it was an upgrade...you're obviously gonna have to take it... you're gonna have to learn it...there is gonna be people in that job that will know it and will be able to help you.

This desire to do what she sees as necessary frames much of her current experience.

> I'm sitting there and I'm bored but I know I've got to go there...I've got to pass my maths.

She is aware that she can take a long time to grasp an idea or achieve competence in a topic, but, however, difficult, she somehow maintains her effort:

> and sometimes I sit on my own and do it...I have to keep ...to read it over and over again. And further: (I get there) in the end...but it takes me a long time... see that decimals took me about two years.

However, this doesn't mean that all of her motivational effort is focussed entirely on the end goal. That would be difficult to sustain for a young person whose recent experience of mathematics is one mainly of failure. At college she encounters a regime different to school, and one that has elements of motivational satisfaction that sustains her. And her experience at college is different to school:

> It's not just M (teacher)...working in groups...we all laugh about things... she's quite laid back...she lets us just talk if were all doing the work...most teachers don't let you do that...they like you in silence.

An important point here is that there is sometimes an implication in research literature that low-achieving and disaffected students lack metacognitive skills such as effort and persistence. On the contrary, Helen is typical of many students in this sample who have to demonstrate a range of motivational and behavioural resources in the face of quite severe difficulty.

> like I'd sit there and if I'm in my own little world...I'll sit there and do it...I came out 100%...so...I think...I've got to concentrate loads as well.

She is also quite self-aware of how she learns: “I have to keep ...to read it over and over again.”; and later: “I think I have to learn it for ages.”

Her relationship to mathematics represents what we might call a utilitarian contract. From the evidence here, we can interpret the ‘terms’ of this contract from Helen’s point of view as: it’s important to get maths; it will be useful to me; if you give me confidence I can do it and if it’s not too painful; then I will try and try. Many of the students in this sample have settled for a similar version of such a contract.

THE CASE OF EVE

Eve is a bright and articulate interviewee. Her performance and attitudes to mathematics are in stark contrast to her achievement in other subjects (where she is undertaking A levels). Her narrative is suffused with expressions of negative emotion in her relationship with mathematics. She was articulate, being able to describe, mainly in the first person, her distress in relation to learning maths.

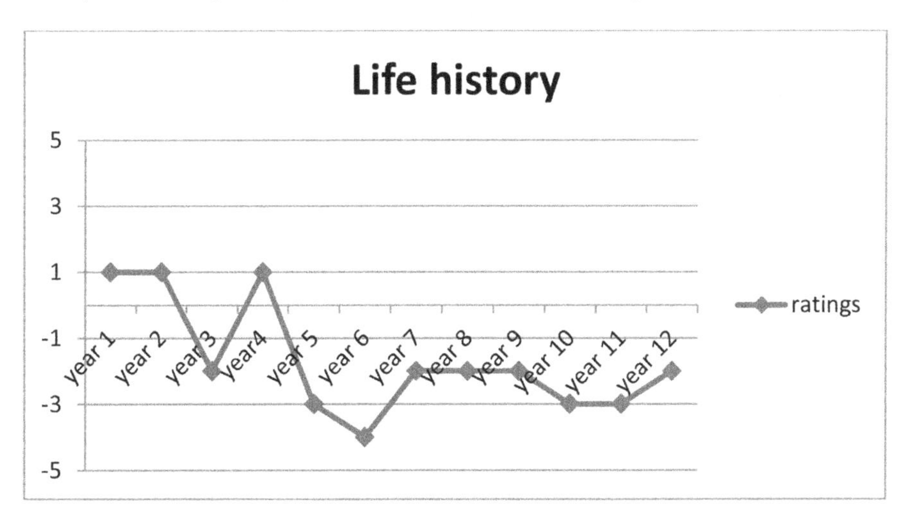

Figure 8. Eve's life history

Her mathematical life history tells the story quite vividly. She starts low (+1), but this is the highest it ever gets. She suffers a dip (to –1) in year 3, a slight recovery to the positive in year 4, but thereafter it goes very negative, and although it increases slightly, it never again recovers to the positive. It can be concluded that Eve descended into disaffection quite early in her school career. She is quite stressed about maths (scoring 4 on TESI-ME for stress), and her strongest reported negative emotion is anxiety (score = 5). An analysis of her narrative initially gave 23 separate codes (although there is some overlap), and these were supplemented by a range of interpretative annotations added by me.

Although maths is a source of serious negative affect for her, she manages her involvement with the subject, and looks likely to see it through to a reasonable outcome (achievement of a GCSE 'pass' grade). What characterises her narrative is that these negative emotions are never very far from her consciousness in her experience of maths, although she has learned to keep the worst of them under control. This struggle between panic and control is a key theme.

Decline into Disaffection

By her own account, her early experiences were that maths was simple and she could do it, until she suffered a significant dip in year 3. She ascribes this to a change in teaching strategy – moving from using equipment, physical objects, and diagrammatic representations, to examples on the board. As she says, this style "brought me down a bit", and causes her to be confused. She concludes: "I knew I couldn't learn like that", and she lost interest.

She has a strong awareness of her own learning style, and understands, even at that age, that she cannot learn by just looking at a board. And although later she was offered help, by then, in her words, she had "lost track", she was getting confused, and she became aware that she was behind everyone else. She is also becoming aware that she is being labelled as problematic. She talks of being pushed down. She notices that teachers are focussing on those who 'get' it, and she is being ignored, and the teacher didn't want to help her.

Her life history shows that Eve's performance and attitude did stabilise and improve somewhat in year 4 – which she attributes to a new teacher. But slowly, through year 5, she fell further behind.

> but obviously I'd lost track of everything…because the things I was learning…were things really far behind everybody else…I just ended up getting really confused…just lost complete track of everything I was meant to be doing…

And although her struggle had been noticed, the way it was dealt with just compounded the problem:

> E – and obviously the consistency of the teacher…you know…helping those who are getting stuck…more…because I felt myself constantly being pushed down…when I felt I should have been pushed up...
>
> Gl – how do you mean being pushed down?
>
> E – like…I have noticed that a few of.my teachers would focus a lot more on the ones who were getting it rather than the ones who weren't…and I found that a bit strange...
>
> Gl – so you were getting ignored?
>
> E – yeh… which was really strange and that's why I felt like I just got dumped on somebody else…because the actual teacher didn't want to help and wanted to push the other students further…

Gl – it's a terrible feeling…

E – …it's a horrible feeling…that is definitely my confidence in maths gone… and it was definitely from a young age…it's just gone…

For Eve, this creates a climate of severe anxiety, and as she says: "It's all the way through then."

The Experience of Disaffection – The Broken Frame

Taking all of this evidence together at this stage, we see:

- She knows she doesn't learn well in the way the subject is being taught.
- She feels she has 'lost track', and is confused.
- She is aware she is behind others in the class.
- She feels the teacher has no time for her.
- Her confidence has gone.

The sense of competence, confidence, enjoyment and trust is now absent. There is no sense of safety – nothing to protect her from exposure to anxiety and humiliation. Her protective frame around mathematics is now broken. She is highly disaffected.

Her feelings of powerlessness and hopelessness have two powerful connected consequences. One is the way she views herself adversely in relation to others. The other half of this equation is how she is viewed by others – especially teachers. At an emotional and transactional level, she feels what she perceives as their lack of patience with her. She feels that she is being labelled from an early age, and thus being treated as 'dumb' by both teachers and other pupils, and this is clearly important to her. She recounts having been put into a low performing group at year 3:

E – yeh…and…even though they wouldn't say to you…this is an ability group…you'd sort of work it out for yourself…easily…it's just…obvious isn't it…it's just completely obvious…and obviously…when you pick up on that…and you notice you've got the teacher at the front of the class…y'know going round to all these…like.. really able students that can do it…and then the teacher comes to you and its sort of like…you can see their frustration at having to explain it over and over again to you…and that's something where it knocks your confidence down… you sit there and you…uhh… don't wanna ask…because you don't want to get frustrated that I can't get it…and that's how I always constantly felt that I couldn't ask because…of their frustration at having to explain it to me again…

Gl – and you're that kid and you know you're in that bottom group...how does it feel?

> E – it's horrible…especially if the teacher asks you a question…and says 'eh Eve…what's this..?' and all these able groups turn round and look at you and like…she's not gonna get it cos she's in a bottom group sort of thing… so…it's definitely one of the worst things I've ever had to go thro in maths…is being put in ability groups…

The labelling also works in subtle ways – even when she is offered help, the help comes with a symbolic, and very public message:

> and I don't think it should be the kind of help …where …unless really necessary…they should be pushed on to another teacher…to be taken outside of a class…to do it then…because…I feel that makes you even more insecure about maths…it's sort of like why can't I just do it within the classroom…

Eve appears to be acutely aware of the public nature of her lack of competence and what she perceives to be the adverse judgements of others. This compounds her lack of self-confidence, and creates a kind of multiplier effect. And what she perceives to be any public display of her incompetence she feels acutely:

> and everybody's like…Eve's never…you don't wanna be that person that's never been on there so…it's that pressure there...and I don't think I could handle pressure…with a subject that I can't do as well.

Eve's use of language is instructive – her lexicon of disaffection includes the following:

> *Horrible embarrassed anger terrifying frustration pressure panic confused boredom*

Not surprisingly, this makes the maths classroom a frightening place for her. This passage captures a number of features:

> I just remember every time it would be maths it's just like…anxiety would be horrible… because …it's…it's…it's embarrassing…you know… if a teacher says 'can you tell me what this is?' …you're sat there completely blank even tho you're concentrating your hardest…and they ask you a question and you're like uuurr…and your whole class knows it and they're facing on you and you don't know it…you tend to feel really embarrassed…

Here we see that being in a maths class induces anxiety, and as we see above, this is variously expressed as terrifying, or panic. This indicates that she is in a serious-conforming state when arousal is high. The situation is perceived as threatening – or in her words, 'horrible'. But she also mentions the word embarrassing here (as elsewhere). The emotion derives from self-mastery, losing, and is close to humiliation. The comparison between her own sense of mastery, and the comparison with others is damaging and exacerbates the intensity of the emotion.

As well as these in-the-moment experiences, there are more pervading or global influences on her experience. For instance, her sense of hopelessness and powerlessness is expressed thus:

> yeh…at that point…even tho I was ill…I sort of slacked more in it… because I thought even if I was well…I wouldn't pass it anyway

Eve's highest scoring negative emotion on the TESI-ME was anger. Here is how she describes how it comes about:

> I think it's sheer frustration…you know when you're doing a topic and… you somehow can't seem to grasp it…or you want to grasp it…like we did pi charts…the other week…and I wanted to so badly to get it...and I knew that I could get it…but my whole attention was just going all over… because I was focussing so much on trying to get it…rather than just actually getting it…and I think it's that frustration that really annoys me…when you're trying to do something so you can't get it…

This is a good description of a reversal. We have seen that her 'default' setting when faced with mathematics in the classroom is often anxiety, which suggest that she is in the serious conforming state combination. She wants to do the task – 'you want to grasp it…you want so badly to get it'. However, despite her efforts ('my whole attention' and 'focussing so much') no progress is being made. This leads to a build up of frustration, and this is a known cause of reversals. In Eve's case she reverses from conforming to rebelliousness, and this leads to a switch of emotion from anxiety to anger.

> Gl – so when you're stuck in that anger...what do you do? How do you deal with it?
>
> E – I end up going back to being bored… because I'll just sit there…I'll just be like uuugh… because if I can't do it I'll just give up on myself…sometimes…
>
> Gl – because it's better than being angry?
>
> E – yeh

How does she get back to being bored? Boredom occurs in the playful-conforming state combination with arousal low. So some psychological changes need to take place to achieve this. In fact the 'uuuugh' suggests that the operative emotion might also be sullenness, which suggests that she adopts an 'I don't care' stance, reverses from serious to playful, the arousal lowers and the anger becomes sullenness.

The Experience of Positive Affect

Despite the strength and depth of her disaffection, like so many others reported in this study, Eve's experience of mathematics is punctuated by islands of positive affect – experiences which create a sense of satisfaction for her. These occur when she is able to satisfy the needs of her motivational state at that time. Also, like many others in this study, one of her sources of satisfaction is to be able to help others.

> E – I do like to help other people…but not…but obviously help them in a way that I would like to be helped…
>
> Gl – which is how?
>
> E – which is obviously starting from how I'd work something out…and not say'... its this'… you need to do this'…like actually show them…like write it down for them and show them…so they can see it…rather than me just being like you have to do this...and let them write it down…cos if they don't know how to do it they're not going to know how to write it down…(long pause).

This isn't the only potential source of satisfaction for her. That also comes from the achievements and landmarks along the way. These range from the fine-grained (achieving a task or right answer) up to more global outcomes (passing her level one).

Not all topics hold the same sense of fear for Eve, and she is able to use her resources and effort to make progress. She talks of a sense of relaxation and pride when she is able to do something independently. However, these experiences seem to be few and far between.

And so to College – Recovering the Broken Frame

Given her relationship to mathematics, it is perhaps surprising that Eve may even though have stood a good chance of gaining an acceptable result at her GCSE. Contingent factors at home (illness) seemed to have got in the way of her preparations. Not surprisingly, her sense of powerlessness and helplessness meant that she was resigned not to do well. As she says: "at that point…even though I was ill…I sort of slacked off more in it…because I thought even if I was well…I wouldn't pass anyway."

But coming to College held no fear – on the contrary:

> I'd probably say…I know I was quite relaxed about doing maths at college… because it is that thing there where...if there's students who can't get it at school…they know they're gonna try and get it at college…and…so I knew I wasn't going to be the only person who was going to be stuck… because it was going to be the whole class that was there.

This notion of 'I won't be the only one', so I won't feel dumb in front of the others is an important part of the restoration of her emerging protective frame.

Eve's way of coping with stress is also not to try too hard. The message seems to be 'if I don't invest too much, or try too hard, I won't get hurt.'

> …in comparison to all my other subjects… because I feel I'm not focussing on it more…I'd say it wasn't as stressful…but if I was to put 100% effort into it…I think it would be very stressful…

The other strategy that seems to work for Eve is repetition. She achieves success in the Level 1, although being in a level 1 group is in itself a humiliation for a girl as bright as Eve, who is doing A levels in her other subjects. We can get a sense of her approach from the following story:

> I think the thing is within a job…is… because it's something you're doing every single day…and it'll be the same type of maths you're doing every single day…you kind of grasp it… I mean I have a part time job at the moment and working in a shop…you're having to use a till…that terrified me at the start…oh my gosh what happens if I can't give the right amount of change …I get embarrassed if you don't give the right amount of change…and that kind of thing…and then I thought…if you keep doing it more… I learn it more… and because I do it so often now...now my confidence is up there so...it's much more.

It is interesting to note how the fear of failure and embarrassment drives her need for competence – even though she is 'terrified'. She has to go through the panic and work very hard. Repetition seems to be the key for Eve, together with a motivational need and rationale for doing what's required. In the face of these difficulties she shows determination and persistence – keep doing it more and I learn it more. And as she does it, she learns, and as her competence grows – so does her confidence.

In the literature there is evidence that low attaining and disaffected students lack motivation, and also lack the metacognitive skills and resources to deal with challenging situations. This is clearly not the case with Eve. She demonstrates at times very strong drive to improve her performance, and a range of skills and personal resources to help her to do this.

> like we did pi charts…the other week…and I wanted to so badly to get it…and I knew that I could get it…but my whole attention was just going all over… because I was focussing so much on trying to get it…rather than just actually getting it.

She has enough self-awareness and sense of agency to be able to make progress:

> I mean… the teaching I had last year wasn't as good…but I got to the point where I needed to focus…I'd sort of come to the realisation that I need to do this for myself because I'm not going to get any help

> right now…due to my (?) and missing my first unit…if I can put in 100% effort I think I could easily pass…but that's obviously down to the way that I use my time

Like other students in this study, Eve has settled for her own version of the utilitarian contract. In Eve's case, this has a paradoxical nature – I want to pass my GCSE mathematics so I never have to do maths again. Eve has developed a maturity about her own learning that enables her to overcome the emotional challenge and burden of doing maths that enables her attempt this. However, it is also clear from her own account that she needs help and support in order to do this.

THE CASE OF MEENA

Meena appeared to be a very mature and articulate subject. She contributed enthusiastically to the interview. Her life history gives no hint of disaffection in the first eight years of her school career (she scored each of those years +5). There is a slight decline of 1 point which maintains for 3 years, and she then declines steeply into negative territory for year 11 and 12. Despite having appeared to enjoy maths at school, she failed GCSE (gaining an 'F'). In terms of her current situation, she appears to be highly stressed (scoring 7 for overall stress on TESI-ME), with scores of 7 for 6 of the eight stressors. From happiness to deep despair in a few short years?

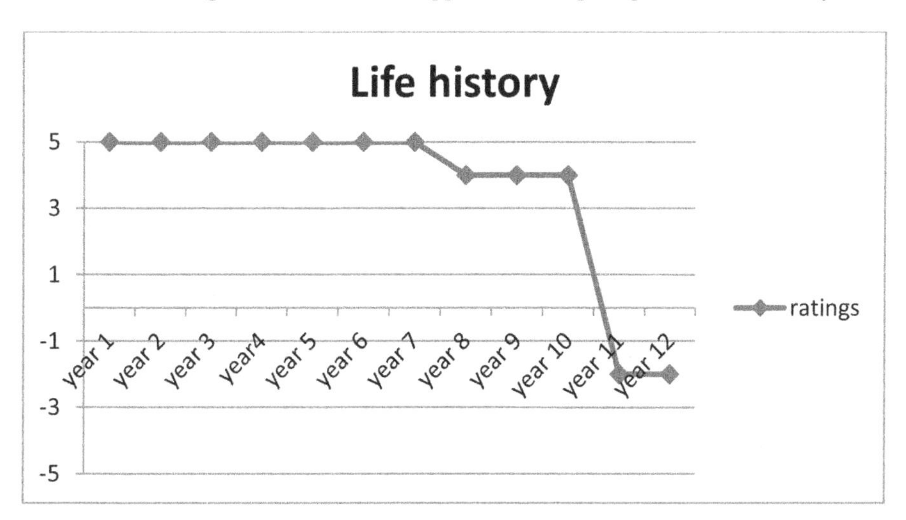

Figure 9. Meena's life history

The First Eight Years – All Is Well, but Perhaps Not!

Although the balance of the interview focusses on her decline in affect, and the current situation, there is evidence of her enjoyment of maths in her early years. As she says:

> yeh…when I was younger and in high school. I was really good at my maths

Later, she gave some clues as to why this was:

> Because it was fun…it was different…but even tho it was like cutting up and sticking…whatever you're doing…colouring…matching things up…you're still…I tend to learn more from that than if you're giving me a lecture… because as you're doing it you're still learning thro what you're doing…

Not for the first time, we see students enjoying the active and materials-based approach of primary school. Students seem to understand the importance of 'learning by doing' and find the exposition-focussed approach of later years of schooling much less effective. Meena mentions, with a hint of disdain, "giving me a lecture", and this sentiment is echoed in a number of narratives in this study.

Although she marks this time as high, which suggests that her memories are of mathematics as unproblematic and even enjoyable at times, it is tempting to wonder if all was as well as she claims, since her decline was so delayed, so rapid, and so deep. Perhaps the seeds of her later poor performance, and collateral decline into disaffection can be seen in a number of ways. By her own account, she had a tendency to be easily distracted.

> M – yeah…I get distracted quite easily…
>
> GL – you're not alone in that…how is maths compared to other subjects? Is it easier to get distracted in maths than in other subjects?
>
> M – mmm yeh…it's because I don't like it…that's probably why I let myself get distracted…I don't know…like I just forget everything…because when…I lose my concentration…I notice that as you get older you lose your concentration… you get distracted more easily…you lose your concentration… You don't end up listening or something like that…it depends what environment you're in… like the atmosphere as well cos if it's a quiet classroom I tend to focus more… just listen…but if everyone's talking or there's muttering whatever… I tend to just listen and do my work…I probably gossip and end up not doing my work in the end…

This reference to concentration is important, because, for Meena, like for many others, it is the key to her ability not just to perform a task in a procedural manner, but to embedding knowledge in a more meaningful way.

> I need to understand like the question…how to do it…cos if you tell me once…I'll probably get it right and do it…but if you ask me again I probably wouldn't know what to do… because you've only explained it once and with me… I need to explain it over and over again for me to understand what I have to do otherwise I just forget it…and…(laughs)… I can't do it.

Although she is compliant – she will do what is asked of her, nevertheless she finds mathematics difficult, and thus does not like it. And this difficulty comes from the need for explanation and a degree of repetition:

> It's just some topics that like...I'm not really ... because with me... I kinda like...if you tell me how to do it I'll listen and everything but it won't go thro straight away...you'll have to tell me quite a lot of times...for me to understand.

Despite her high scores, it is possible to see that whilst she coped with mathematics (she was placed in a middle group), both her attitudes and her performance suggest that all may not have been as positive as appears in the life history.

Decline into Disaffection

Her decline began, as she puts it – "coming on to the big school". In the local authority in which she lives, pupils are transferred at the start of year 10.[1] However, it is interesting to note that she charts her decline as from year 9. Although we have no direct evidence, it is likely that there is an increase in seriousness and difficulty associated with this transition, and test/exam scores will have been a part of this.

Meena's narrative related to her decline is dominated by the theme of exams, although her description makes it clear that this includes in-school testing of any sort. As she says: "It got worser in maths."

> I kinda get nervous...mmm...I kinda panic a bit which I shouldn't... because I have panic attacks...so I shouldn't really worry like. (??)

And when asked what causes this panic, she replied:

> because it's only maths though... because maths is important...I know it's something that you need in life...you need a good ...like...maths grade...cos everywhere now involves maths...depending on which work you wanna do... it involves it so I know it's very important...so that's probably why I get really nervous cos I want to do well cos I know it's very important to have...

Like other students in this study, Meena battles to understand why exams are so different, and she has difficulty understanding why she can't perform in test/exam conditions, when she can do it in the class.

> but it's because the exam is very different to what you learn in the classroom... like the wording is different...the methods may be different...so that's probably why I don't understand the paper... because I always think to myself at the end of the paper because I've failed...why did I fail?...I know this... I go thro it in my head...it's just not...

She got an F (fail) at GCSE, and that appears to be a surprise as well as a disappointment to her, and, according to her, also to her teachers. And this exam anxiety and associated exam incompetence is a real phenomenon.

> I felt disappointed...even my teachers were disappointed... because I do so well in my lessons...and all my coursework and my classwork...I don't hardly ask for help in classroom... I understand the work...then when it comes to the

> actual exam it's a fail for me and my teachers think where did I go wrong… how come you failed…when you shouldn't have failed.

It is easy to understand her sense of confusion and unfairness when she compares her performance in the classroom with her exam results. The 'where did I go wrong' is a clear cri de coeur. She carried her anxiety and panic with exams through to her college career, and she fails her attempts at a re-take.

> no…I got worser in my maths…yeh, I got worse…and it's kinda pressure as well now…so because like…this is a re-take…for gcse…and I've done it like twice now…and it's just pressure… because I'm behind in uni as it is and I want to go to uni…so its pressure on me…

The Experience of Disaffection

Her failure in exams seems to tip the balance in her perception of her affective experience of mathematics. Her panic amplifies over the years.

> I don't like it (laughs)…noooo… I find it difficult…hard to understand…

> me being stressful makes me worry a lot…like…it just makes me worry a lot…and when the exam comes up…maths and that… because I know it's important and I need to do well in this…I tend to not sleep…I tend to not eat… because I worry a lot…the doctor told me I shouldn't like … because I used to be ill previously…and I have panic attacks a lot…and the doctor tells me I worry too much and I should avoid things I don't want to talk about…should like concentrate and don't worry…but (laughs) but I still worry and things like that…

Not surprisingly, this sense of panic begins to pervade her whole experience of mathematics at college. She is disaffected and now appears to become more acutely aware of the negatives in her experience. One aspect of her disaffection is boredom. Interestingly, the following articulation of her boredom goes back to her school days, when she reported a strong positive relationship with maths.

> no… I just don't find it interesting…like…the way that the teachers explain it…it's like giving a lecture…and when someone gives a lecture…you tend not to listen… because its just one tone of voice and you're just going on and on… and I just tend not to listen…like if you're doing a demonstration or change your voice tone.

Another, perhaps predicatable aspect of her disaffection is a sense of humiliation.

> – it makes me feel like I'm dumb and stupid because of my grade…and I feel like…feel like…when people go…ah…what you doin at college and you go maths gcse or whatever…I'm ashamed to say I'm doing it because maths… is like very important…and I just feel dumb and stupid…and if someone else

> knows more or got a better grade…I just feel like ohmigod I dont want to say mine…

As with others in this study, the humiliation of not being able to do something, or even of having failed, is compounded by her perception of what others will think of her.

There is evidence that Meena is conforming, in that she not only wants to do well, but tries hard in class to do what is expected of her, and what she expects of herself. There is also evidence that in terms of her behaviour and in terms of classroom outcomes, she does as well now as she did earlier in her school career. And yet, her affective landscape is clearly massively different. Why is this? The answer seems to be that her experience of maths has been dominated by testing regimes since year 9. This alone might be enough to colour ones emotional and motivational orientation to the subject, but in addition, Meena has to carry her history of failure, and the impact on her sense of self, and her anxiety/panic in relation to forthcoming exam requirements colours her entire experience of maths.

A Sense of Satisfaction – I Feel Clever Even Though I Feel Stupid

Despite the analysis given above, and like so many students in this study, Meena manages to maintain a sense of motivation. What is interesting is that not only does she maintain a sense of utilitarian purpose in relation to mathematics, she seems to value learning, improving and achieving for its own sake.

> yeh… I also want to improve it anyway…just like…even if I don't need it I still want to improve it anyway because it is very important.

Her performance and progress in class do bring her some fleeting sense of motivational satisfaction.

> M – yeh…I would say I'm quite modest…in my classroom when I do my work…if Ido my work…like it's all correct…I tend to… people tend to copy my work…so that kinda makes me feel happy…it makes me feel clever…even tho I feel stupid…
>
> Gl – it makes me feel clever…even tho I feel stupid…how do those work together?
>
> M – it makes me feel clever because someones taken the interest to look at my work…to take information…but then again…I feel stupid because of my grade…so…that's on one side…

The experience of feeling clever versus feeling stupid is an interesting contradiction for Meena. It appears that the feeling stupid is a more pervasive and stable condition, and the feeling clever is more transient and volatile – it doesn't last, and it doesn't override the sense of feeling stupid for long.

Another aspect of Meena's motivational make-up and style concerns arousal-seeking. We have already noted her propensity to be distracted. She appears to spend time being bored and needing distraction. Here she is talking about her choice of the card 'mischievious':

M – when I don't understand something…and I can't do it…I just give up… and I'll talk and gossip and I'll probably go on my phone…I'll do something else to occupy me, then rather trying to do the work…

G – do something else…anything else?

M – yeh…it would be even doodling…or sitting there just staring out the window…

G – what turns the mischievious on…what flicks the switch? because you feel what?…then get distracted and want to do something else…

M – like if I can't do the work or someone's talking to me…yeh…if I don't understand something…I'll tend not to do it…

The sequence here is interesting. She is trying to do something in class (serious-conforming). But she doesn't understand, can't do it, and not being able to make progress she is feeling frustrated. This appears to cause her to reverse into the playful-rebellious state combination, which creates a 'what the hell' frame of mind. However, in the playful-rebellious state combination, with low arousal, she will seek arousal and paratelic pleasure ('something else to occupy me'). Meena clearly spends a good deal of time in lessons in the playful state. She reports that she responds well to any activity or pedagogic approach that creates interest, arousal and therefore paratelic pleasure.

M – mmm…having fun…sometimes the lessons can be interesting… like sometimes we go on the laptop…and we'll go on mymaths dot uk or something…and do problems on there…and there's also maths games on there as well.which I find…that tends to help with my work.

Gl – does it work?

M – yeh…if I don't understand a subject…if I see that on the game…if I don't understand a topic…if I see that topic on a game I'll tend to play it because I like to play it and it helps me to understand because I like playing games…it's just a different way…I find…maths games and that…I find it different…it's just a different way of learning…it's all…like the structures different if you know what I'm trying to say…

Gl – sure…and how's that make you feel…compared to normal lessons… lectures?

M – oh… It makes me feel happy

Notice how she is able occasionally to turn a situation of potential telic disaffection (don't understand a subject) into a paratelic arousal-seeking opportunity (by making it a game). She is managing her own perception of the situation to gain paratelic pleasure and satisfaction.
Another source of motivational satisfaction for Meena is helping others.

> yeh…helping others…like certain topics that we do in classroom…I'm not so good at…I tend to ask for help if I try to understand it…but if others don't… the people who are sitting next to me…if they dont understand…sometimes I wont let them just copy my work…I'll explain it to them in a different way that…the teacher has,,just to break it down…so they understand…and it does actually work…like one girl…the girl who's sitting next to me right now… she's the one who don't want to copy my work no more…she'll ask me to explain it rather than the teacher…I'll break it down…make it more simpler… which makes me feel happy…that also it's helping someone…it makes me feel happy (laughs). It also makes me feel quite important as well that she'd rather ask me for help and to explain it more…rather than the teacher

As has been noted with other students in the study, helping others is a motivationally rich context, in that it appears to satisfy a number of motivational needs for her. She gets autic pleasure (feeling virtuous) from the fact of helping others. She gets a sense of self-mastery, winning (power and control, which leads to pride) from the sense of importance that they ask her rather than the teacher, as well as from not letting them just copy her work, but by explaining her own method. She gets a sense of freedom and asserts her individuality (rebelliousness) in doing things in a different way to the teacher.
When asked what her best experience of maths is, she replies:

> M – mmmm (trying to recall)…probably when I concentrate
>
> Gl – what happens when you concentrate?
>
> M – I tend to get more work done…which means because I'm doing more of the same problem but it been put in a different way…I'm understanding it more so when I do it again… I can get it straight away…I know what I'm doing…I'm not sitting there thinking oh how do I do this now.
>
> Meena also has an interesting interpretation of what makes people be 'clever' at mathematics.
>
> Gl – were they born clever?
>
> M – no not really…everyone's born the same…I suppose it's the environment… …I think it's the environment and the atmosphere…that makes you clever.

When asked what makes the right environment, she goes on the say:

> maybe an interesting lesson… a nice environment to be in…so how the classroom is looking…(pause) not too much noise…but then not totally quiet…

> just a calm relaxed place where you can go and sit there and feel comfortable… and your mind is just free to…

The ideal environment also includes opportunity for paratelic pleasure, fun and interest:

> well, we do work in groups which is nice… because it's activity work…it's something interesting to do…I find it quite fun at times as well… because you're socialising…makes it more better…I think it makes it more better… like easy to understand… because you can share your ideas and your thoughts of how to do… how to work out a problem…and you can find the best solution for your self.

In terms of her motivational style, a great deal can be inferred from the evidence here:

- She spends a good deal of her time in the serious state.
- When she is in the serious state, she manages classroom tasks and activities well. She can concentrate and she can apply effort to achieve short term goals. She appears to have a mature self-awareness to enable her to understand the environmental conditions under which she can learn.
- She is mostly compliant.

Despite spending a good deal of her time in the serious state, she also has a strong playful side and a need for arousal – this explains her boredom, and her susceptibility to distraction

In terms of transactional states, self-mastery (losing) seems to be the default mode (although we can speculate that she experienced 'winning' much more in her earlier school career). Even small wins seem to be temporary and don't seem to override her global self-perception of herself as 'stupid' at maths.

NOTE

[1] In this local authority, pupils are in High school until year 9, and transfer to Upper school at year 10.

THE CASE OF RAZ

Raz was quite a serious-minded subject. Although his response to the TESI-ME suggested he was not unduly stressed about maths, his life history showed a deep disaffection with maths at stages in his school career. We can tell his story in a number of phases:

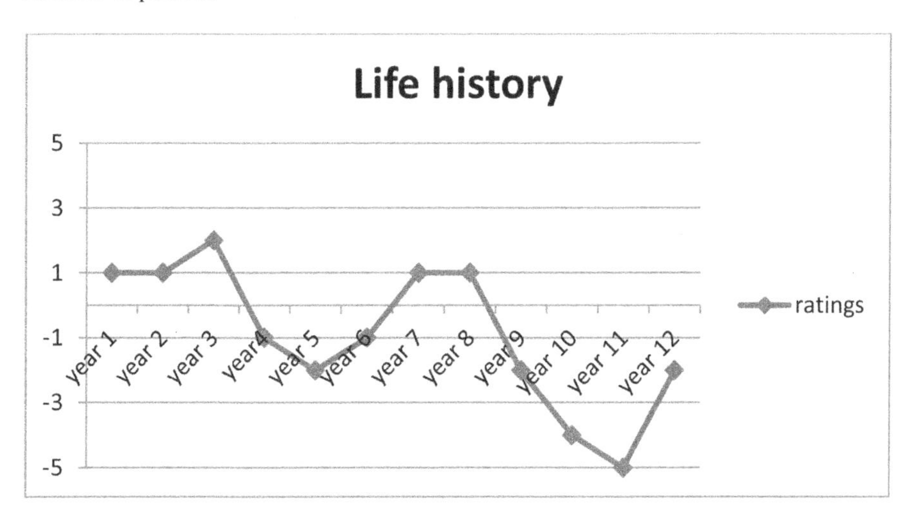

Figure 10. Raz's life history

The Early Years

His life history shows that Raz was quite positively oriented to maths in the early years. He could do it, and it was unproblematic.

> I had a rough understanding…of maths in year 1…which carried on to year 2 cos it was mainly the basics…just addition and subtraction which…like…I didn't find it hard to do.

Not only could he do it, he seems to have enjoyed it at times, and this enjoyment corresponded with a sense of learning something new.

> yeh…it wasn't really that challenging so I didn't mind doing it…year 3 I think it was multiples…like times and divideI actually enjoyed that bit cos it was

> different…cos I spent like two years doing adding and subtracts…so it was something new

It is interesting to note that he remembers fondly the style of mathematics at that time, involving as it did, the use of games and equipment, and involving the energy of the pupils.

> yeh it worked…we got like a fun exercise out of it…and we got multiples… times stuff…and we had to run and grab the right number and stuff…we learnt all that stuff.

He contrasts this style of learning to that of secondary school (see below), which he sees as very different.

Decline into Disaffection

His early experiences are in contrast to his later view and experiences, when mathematics becomes more dull, and more of a chore. As he says: "I just kinda dropped…cos we started doing different stuff." This different stuff was fractions, which he didn't enjoy, and didn't 'get'.

And as the mathematics changes and becomes more difficult, so his enjoyment of it lessens. This accounts for the dip in year 5. This pattern repeats for Raz: that the requirements of the maths gets more difficult, and this causes a dip. On the other hand, as will be seen later, he then works at it and gains proficiency, and there is a jump in year 7. However, after primary school the opportunities for satisfaction change qualitatively. Now, maths becomes boring and dull.

> …I guess it's coming to secondary school…that's when maths actually starts getting a bit duller…cos it's more just…do the work…it's that kind of environment…that's why they say maths is boring…because it's just sitting down and doing the work and go…

Perhaps the most significant watershed of his decline is that he fell out with a teacher. The evidence does not suggest that this is primarily to do with the mathematics (although it only happened in maths), but seems to be more to do with personality. However, at this point, Raz withdraws all of his effort.

> yeh…it was just the teacher…it felt like she was just singling me out…like picking on me…so I just wouldn't do anything for that one teacher…

In his own words he does no maths at all, and this accounts for the serious decline in years 9, 10 and 11. At this point he was excluded from mathematics classes. Like so many students in this study, once ground is lost in those vital years, he never fully recovers either interest or competence in the subject. Although he and the teacher reach some kind of compromise agreement just prior to exams, he revises, but, not surprisingly, is unsuccessful at GCSE and gets only a grade E.

The Experience of Disaffection

Raz gives a global judgement about his relationship to maths:

> I've never been like good with maths

His judgement may be coloured by subsequent events, or perhaps he is aware that he has had to work hard to gain competence, and this comes about over months and years, rather than immediately.

He also makes it clear that it is mathematics that is the problem.

> yeh…that's always the thing with maths tho…I've always had to work at it like…and…try and remember it…english came just naturally to me…

He is able to articulate why he thinks this is:

> english…I like that because it's more creative…I got to express myself thro words…science…I kinda thought that was the middle subject between english and maths…

These perceptions play out in his achievements at GCSE:

> my english was ok…science…science was there…I knew quite a bit for science…I had just enough to get a 'C'…and maths…cos obviously…I just didn't like that one teacher…I just chose never to do the work in the lesson…

A number of things have the potential to cause negative emotion for Raz. One is competitiveness. But competitiveness can cut two ways. It can motivate an individual to perform, but it can also de-motivate if comparisons with others are adverse.

> Gl – is it (competitiveness) a good thing?
>
> R – it's a good thing because it kinda pushes you to do a lot better…but at the same time it's kinda like harsh because if they manage to even get like one mark…you're gonna look in your head as one mark…but I still lost to my mate…they're gonna rub it in because they've obviously beaten you

Raz hadn't talked a great deal about negative affect, so he was shown the photo of a boy (apparently upset in a mathematics class).

> Gl – has that (photo) ever been you?
>
> R – at one point…that was like in secondary school… because me and my mates all were in the same group…and we'd always say…like your dumb… I'll do better than you in the maths exam…but when it came to it…one guy just kept on beating us in the maths exam…so we just like…we felt wounded cos it was just…you never did anything…cos we thought…yes…we're smarter than

> him cos we …he was like a new kid…we thought we'd be better than him…so …but it forced us to improve our maths.

In motivational terms, he has shown that he has experience of the negative aspects of a competitive stance (self-mastery, losing), and he seems to manage his competitive need by sitting with Beth and Tasha, since he can then perform better than them (self-mastery winning), and even use this favourable position to teach them in turn.

On the TESI-ME Raz scored quite low for stress.

> I would say maths would be about a '2'…cos M's lessons…they're not very stressful…she gives you the work at your working kind of pace…but towards the…obviously you need to learn it as quick as possible…trying to fit it in…that's when the pressure starts to build up…but it isn't…like a massive difference…it's just you have to work a bit faster…

Raz sees himself as not good at maths, he sees it as dull and therefore doesn't like it most of the time, and the contextual reality is that he has failed his GCSE. In similar positions, other students report higher levels of stress, so what makes Raz different? The key to the answer appears to lie in the way he strategises his involvement with mathematics. A number of his comments suggest that he seeks to keep the level of arousal low. He does this because he is most often in the serious state (exhibiting very little evidence of paratelic enjoyment in maths). His description of his respect for M's (teacher) agenda, suggest he approaches work also in the conforming state. In this state combination, high arousal will be experienced as anxiety, and there is evidence of his attempts to avoid this, and indeed, he says this quite explicitly. On the other hand he has learned that he can conquer distress with time and effort.

The Pleasures of Maths

Raz shows evidence of enjoyment and different kinds of satisfaction in maths throughout his school career. However, the qualitative nature of this satisfaction changes significantly with age, and with his developing sense of what mathematics is. In the serious-conforming state that characterises much of his involvement in maths, combined with effort, it is not surprising that he achieves results at times, and these occasions give him a sense of pride in his achievement.

> when I've like…cos you obviously learn multiple stuff in maths…when I've actually finally learnt it and got the hang of it…and I know I'm gonna remember it…I actually can feel a little bit proud…like I've finally got something that I didnt get like…it took me ages to get.

He also gets pleasure from using his sense of competence in helping others:

R – like if I've learnt something and I know I can do it…I don't mind helping other people…who can't…who might just make a few mistakes…cos that's what I do anyway (laughs)…

Gl – but whats in it for you?

R – it just helps me memorise it even more I guess…cos if I've helped them and they've got it right…I know I'll remember it…

Gl – but do you like the fact that you've helped them as well?

R – yeh…I like the fact that I've helped them…that's why I put 'feeling part of the group'…

Gl – so it's related to that…so who do you help?

R – I mostly help tasha…the girl who was sitting next to me…and bethany every now and then…cos with bethany she just forgets the methods…she gets the right working out…tasha's the opposite way round…it's the wrong working out but the right method…

We see from his narrative that working hard to learn something, and the resulting sense of competence is the cornerstone of his positive experiences in maths. It provides a motivationally rich experience which satisfies a number of motivational needs. We have already seen how it gives him a sense of pride. Raz also chose the card 'sense of achievement'.

yeh…I did the sense of achievement one…cos I paired it up with the sense of duty as well…cos I'm doing obviously what is expected of me to do…cos thats M's lessons and she's got like a lesson plan…sense of achievement is like…after I've done it…cos M could give us something really hard to do…to see who can understand or get a rough method of how to do it…and that's my sense of achievement…if I know I've got it right…with my first go…or bits of it right…I feel happy …cos at least I got something right…

This segment also seems to confirm that being able to 'do it' not only generates a sense of personal achievement, but also gives 'sense of duty' satisfaction, which is motivationally about conformity or doing what is expected of one. It can be inferred that the ability to 'do it' also provides the basis for the satisfactions he gains in helping others. There is also evidence here that these satisfactions are enhanced by his sense of working hard and overcoming difficulty in order to gain the satisfactions reported above. Because it doesn't come easily, the satisfaction is all the sweeter.

yeh…at the start of the year she gave us a sheet called bricklaying…and that really did confuse me…but I got that one section of it right…just cos it was related to adding up bricks and stuff…and I at least got something right.

Raz will invest this effort if his motivation has been stimulated by some sense of interest in the task:

> multiples…like I did them when I was younger…but the only reason I got the hang of them was…(X) my teacher put them out…cos if you write 10 and 'X'…and whatever number…it…equals…well my teacher wrote it ten… number underneath with a line…and did it that way…it's just interesting how he laid it out…cos it just took my interest…(laughs)

Raz has enough self-awareness to know that he doesn't always 'get' things quickly. This can cause him confusion, or, in his words: "…depressed because I didn't know how to do it." But it also seems that he has learnt that he can negate the negative emotion and gain a sense of competence by effort and other self-regulatory behaviour.

> …but then I'd go off and revise it…cos like I revised fractions and decimals for like for so long…and I still don't actually get it…I think it's just…I dunno… it's just how they are…it just doesn't stick into my head…

But time and hard work improves performance, so he gets there in the end.

yeh…but if prefer when I'm working to just be on my own…to just learn from my mistakes…as long as someone's marking it to show where my mistakes were…

He will do this, even when the time horizon is long:

> fractions…and decimals…they're the things that I never understood at that time…and it just got worse towards year 5…(nervous laugh) eventually… towards the end of year 5…I actually kinda like started to pick it up… that's why in year 6…I improved on it…

He articulates this approach in his description of the photo of an (apparently) distressed boy in a maths class:

> he probably feels dumb because that's the first thing you think in your head… I'm stupid for missing such an easy mark and what not…but then the smartest thing to do after that would be just to go and pick up from where your mistakes are…and try not to let it happen again…

Raz gives a nice description of this notion of progress:

> uuhhhh…it felt pretty much like…it felt more like it was for that age group…cos like…it felt more like…adult…so you know that you're actually progressing…

His use of the word 'adult' suggests an awareness of the maturity of his own approach. He gains the satisfaction from the consequent achievement:

> and this year…I've only just recently got that 'B' in functional skills…so I felt happy with that…

Utilitarian Accommodation

It seems important to emphasise that his sense of achievement and hard work occur in a frame in which he has come to terms with a maths regime which is without paratelic pleasure in doing the maths for its own sake. For instance, when asked if it is possible to make maths interesting, he says:

> R – I think only the persons actually doing maths can manage to turn it into anything creative…you cant…cos everyone's different…
>
> Gl – what do you mean by people who do maths?
>
> R – the people like me and the rest of the people in the class…cos we've all got different hobbies…and stuff like that…we all like have our own interests…noones gonna do the exact same thing when it comes to having fun…cos you can have maths…like going out onto the field or whatever… and just messed around playing football…but doing it just based on maths… but certain people might not like football…they might be into like rugby and it just clashes…just normal things like that…cos everyone's got different interests…

And later on he says:

> (long pause)…I think that for our age group where we're at…now…there's not any way you can really make it interesting…like in all the lessons I listen to music and stuff…and I just get the work done…but…that isn't gonna make the lesson more fun…that's just gonna make it more relaxed and get the work done…probably …a little bit faster…

In effect, he appears to settle for a life of maths without interest, and a strategy to make it as painless as possible. What does sustain his effort is a sense of purpose and utility.

> cos you need maths to…get decent jobs and stuff like that…cos I want to do graphic design…and maths comes into that quite a bit anyway
>
> it's cos like in life I wanna get older and I wanna be successful…so I know I've got to have…(?????)…it might be something like science that really bores me…but I know it's like that one extra GCSE that might help…make the difference to getting in to college and that…cos like at one point when I was in year 9 I never even thought I'd make it into college…iIthought I'd get excluded like every day…but like towards the end I got all my GCSE's…the only…thing I failed is maths…

This sense of utility can be quite specific as well as general and long term.

> I use maths because I know that if I do a picture…that's 1080 pixels times 720 lines that makes up that picture…so if I want that picture to get bigger and bigger I have to keep changing the pixels according to how many lines I want…and that's how maths helps me in my graphics…and it also helps me with scales as well because I know what a centimetre is and how long it's actually gonna come out if I print it on paper.

THE CASE OF MASUD

From his life history, Masud appeared not to be highly disaffected, but his narrative showed that he experiences quite serious negative feelings and emotions in relation to school mathematics. His early years are moderately positive, but he charts a decline from year 5, and it seems clear that, like so many others, he never fully recovers from this. It is not always easy to understand how he expresses himself, but listen carefully and there is a complex affective landscape described by him.

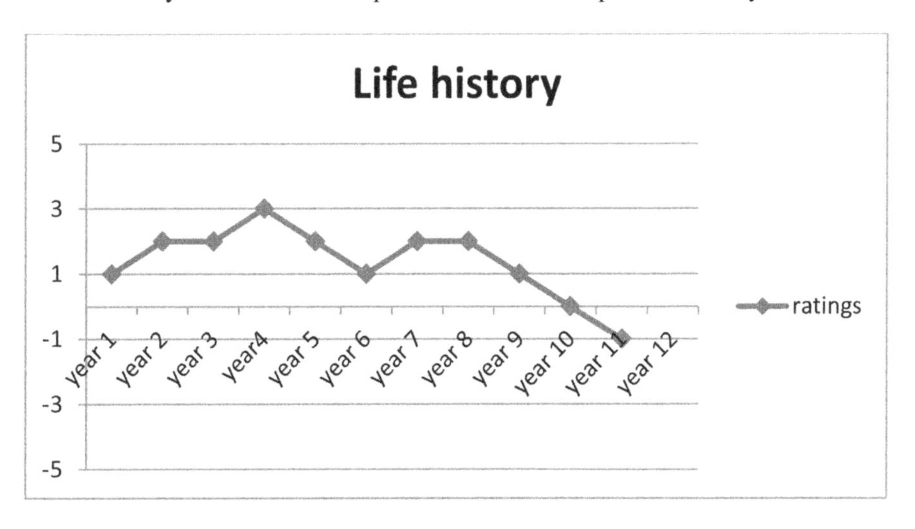

Figure 11. Masud's life history

Decline into Disaffection

On seeing his life history, Masud describes it rather poetically as 'like a rainbow', going up and down as it does.

> I used to counting…like from 1 to 10 then…year 2 was the same thing…and it gradually changed to times tables…in the beginning I was not that good at it…and then gradually I was like good at it…I knew how to do adding and subtraction…and then year 5 I was like one of them students which…on the borderline…and every time I get level 4 and then year 6…I just dropped and I got a level 3…and I knew I failed.

He describes how his involvement in football distracted his attention from maths, but that he kept getting level 4, until somehow it caught up with him:

> M – I dunno…It's like football innit…like sports…I loved sports from year 4 to year 5…I used to play for the team…and when I reached year 5…if I'm getting a 4 every time…I could just mess about and still get the same level… but I never…
>
> Gl – never what?
>
> M – never got level 4 again…I got level 3…that means I failed that Sats…and then I went into year 7 and I knew I failed…and I gradually came back to my year again…and year 8 and year 7 I went to like a religious primary school… then when left year 9 then I couldn't catch up with other… I got moved after that…and then I couldn't catch up.

It also seems to be clear that it is the experience of tests and exams that fixes him into the position of classifying himself (and of others classifying him) as a failure.

> cos I dunno…like at primary school…like a small exam…a small paper…I just didn't like it
>
> no…the problem is…in maths yeh…in class you're doing good…but when it comes to exam…you don't know what happened to you…like you already automatically failed…you don't know the reason behind it…
>
> exams are like…if the teacher gives you work today…to explain it and then that will help you in doing your work again…in exam you get every subject you've been learning throughout the whole year…instead of topic by topic… you'll be getting the whole one in one paper…like different equations and different things to think of

Masud keeps coming back and back to this explanation:

> I dunno…the only time I fall down is because of the exam…and the outcome as such…because in lesson I do quite well…and then I'll be in the top place in lessons and the teacher…they'll be like looking out for me…like the results… but every time I do that exam I just fail…I already know I fail…

It seems from his description that he is searching around for explanations as to his failure, because, ultimately, he is confused by it, and doesn't really know why it should be. The only thing he appears sure about is the difference in performance between class and exams. As seen above, he speculates that having all of the topics in the exam might be a factor. He wonders if the football has taken his eye off the ball, but he can't really understand why he continues to get level 4, until one day, for some reason, he doesn't. Regardless of his performance in tests and exams, he is all too aware of his level, and he is continually aware that he is being rated, and now found wanting.

Since he doesn't understand his failure, at one point, he speculates on other potential attributions for his failure:

> and then sometimes I ask myself…if you're not the teachers pet…if the teacher don't like you will she fail you…that's the bit I don't get…like every time…I can be a clown in the class like…joke about…(?) and can get the teacher angry easily…but every time I do that I ask myself what if…if the teacher hated me and then failed me

This suggests that he has very little trust in the system, or the teacher. What is clear is that this section of the interview is dominated by the words 'failure' (19 mentions) and 'exam' (9 mentions), and the two are clearly linked in his view of the world.

The Experience of Disaffection

It can be inferred from his language that Masud spends most of his time in mathematics lessons in the serious state. In the serious state, focus will be on outcome, and when experienced healthily, a person will be able to describe desired outcomes, although these may be multiple and exist in different timeframes. In Masud's case, the bigger picture, and the longer term timeframe is missing, and satisfaction does not appear to be possible. Experienced positively, the feeling of the serious state should be a sense of significance to our actions and efforts, and progress towards them. Experienced negatively, it will feel like pointlessness. For Masud, his negative experiences have coloured his attitudes and expectations of mathematics. His narrative is dominated by expectations of failure in exams, and he repeats to himself: "What's the point?"

> waste of time…if you know you fail what's the point in learning?

> yeh…plus in maths you get two papers…you get calculator and non-calculator…and even if you pass in calculator and failed in non-calculator…there's no point

At this stage, he appears to have no sense of the potential significance, utility or benefit of doing maths, and this pervades his orientation to it:

> …but…in maths…numbers after numbers after numbers…but you don't know the reason behind it…if you know how to count…what's the use of doing big maths an that?…

"But you don't know the reason behind it" is a highly significant statement in that it gives away that he has a motivational void where a sense of purpose around school maths ought to be.

There are other aspects of negative affect that suffuse Masud's experience of maths. One of these is guilt:

> I can't be happy…you be guilty…cos you know you fail automatically…because you missed a lesson

It is reasonable to assume that guilt here is a transactional emotion, reflecting concern for others. Here is also an interesting expression of his sense of guilt:

> Guilty…because your parents wasted time on you…they want you to come somewhere in life…and time will never come back…every second…and you know you want to pass it…but you don't pass it…and the problem is like… you don't know if it's the teaching ways…you don't know what's the reason behind it

This statement suggests his strong need and desire to pass, together with his serious orientation ('every second…time will never come back…'). But he doesn't pass, and again his expression of powerlessness (self-mastery, losing) at not knowing why. In addition, his negative experience is amplified by his appreciation of how he perceives he has let down his parents: he has received their love and time (which he wasted), and he has not delivered on his part of the 'bargain' by passing, and therefore coming somewhere in life.

It is also not surprising that, since life won't conform to his expectations of it (by passing exams), that he occasionally reverses into rebelliousness and experiences anger.

> because there's a reason behind it…how come I never learn how to do this and everyone in the class can know how to do it

Although we have noted that Masud spends much of his time in the serious state, he reports significant experience of boredom. Boredom is associated with the playful state, when arousal is low. This boredom arises through the experience of repetition in the diet of maths at school.

> that feel like boring…everyday…at the same school

> like in secondary school the same worksheet will come back…every year or every two years…the first one…like a brand new…that will help me think like that…if I see the same paper every two years…I'd be struggling…I'd think like what's the point…if they give me the same paper

Taken together, Masud's experience of pointlessness (having no sense of significance in relation to school maths) and powerlessness (not being able to achieve, exacerbated by not knowing why) create a situation in which it would appear that major motivational satisfactions (serious – significance, conformity – acceptance, and mastery – power and control) are unavailable to him. So despite the relatively moderate pattern presented in his life history, his account of his experiences is dominated by expressions of negative feelings and emotions.

Being Motivated – "I Have To…I Have To…"

Given the depth of his disaffection, it is perhaps surprising to see evidence in his account that Masud is actually highly motivated to achieve something in college.

At one point in the interview, when talking about passing his maths he says "I have to…" This is followed by a quiet pause, the tone changes and he repeats the statement in an emotional, almost defeated manner. Clearly he feels strongly the need to pass, but is presumably reminded of his history of failure – and this weighs heavily on his shoulders. But this serious orientation dominates his approach to maths in college. This works to such an extent that he views enjoying himself as damaging to his goal, and he suffers anxiety if he misses a lesson.

> I think fun/enjoyment – not enjoying my lesson…because you cannot waste time in maths…because every small thing in maths…because if I've missed one lesson in ratio…or (?)…if I've missed one lesson I don't understand that part…I can't find the detail…I can't be happy…you be guilty…cos you know you fail automatically…because you missed a lesson

In pursuit of his goal he shows evidence of self-regulatory skills such as effort.

> I failed in year 6…(?) do my best…so I was trying, trying, trying…when it came to year 9 I came a bit older…a bit mature so I can get something.

> for example…like equations…some equations…you'll never get it…it's so hard…but some equations…it's hard but you know you get it in the end… by the knowledge…I've learnt how to do it…that's the reason behind it…will take you to the answer…I'll be positive like… you've done it by yourself

Because he is motivated, he engages, and because he engages and gains some success along the way, he experiences some sense of satisfaction.

> but proud is when you know you've passed…like if you … if the teacher likes you…you'll be proud like that…I've got their attention

> excitement comes to me like…when… I pass…like in mock exams in lessons… that's when I feel excited…and when I feel excitement then you feel relaxed… and you're thinking…you do it every time…

It is interesting to note here that the excitement comes only after he knows he has passed the test/exam, and that the feeling is momentary since it leads to an immediate lowering of arousal, and the positive feeling of relaxation. In the serious state, when the anxiety is conquered, the arousal lowers and the feeling reverts to relaxation.

He seems to appreciate doing things by himself and for himself, which gives him a sense of freedom and expression of his individuality (rebelliousness) and mastery:

> because your happy then…you'll be glad because…I hate sharing an' that…I hate people telling me what to do…you'll love something which you know how to get…like if I create a pen today (?) it will because…working by yourself… you get all the positives…by yourself…but if you work in a group…somebody telling you… spoonfeeding all day…you're not doing things by yourself

It seems that in non-exam circumstances he is able to accept a challenge, and enjoys novelty and a challenge (as opposed to the boredom of repetition noted above).

> no…I was enjoying what was new…like I never knew times tables…it was something new to me… I started learning it…I never knew about it …and I liked that…when it came to year 5 it was boring to me…like I didn't care.

He compares school unfavourably with college because he associates much of his experience of school as repetitious, whereas at college he is given tasks that he hasn't come across before, and he reports enjoying these.

There is a contradiction in the evidence here. He clearly has a desperate need to pass the exam, but at the same time he shows a sense of purposelessness: he can't see the point in it. Is he doing it for compliance reasons? Or from a need for status? Or is he motivated away from a feeling of humiliation? It is difficult to tell.

CHAPTER 12

THE CASE OF LIAM

Liam represents a somewhat different case study to the others selected here. Firstly, he is a year 9 pupil in a secondary school. It is felt important that this group is represented in this part of the analysis and results, as it is in other parts. Secondly, and more importantly, he is not strongly disaffected by his own estimation. His life history has rises and falls, but these are modest, and all points remain in positive territory. He claims to like maths and says he gets on with it quite well.

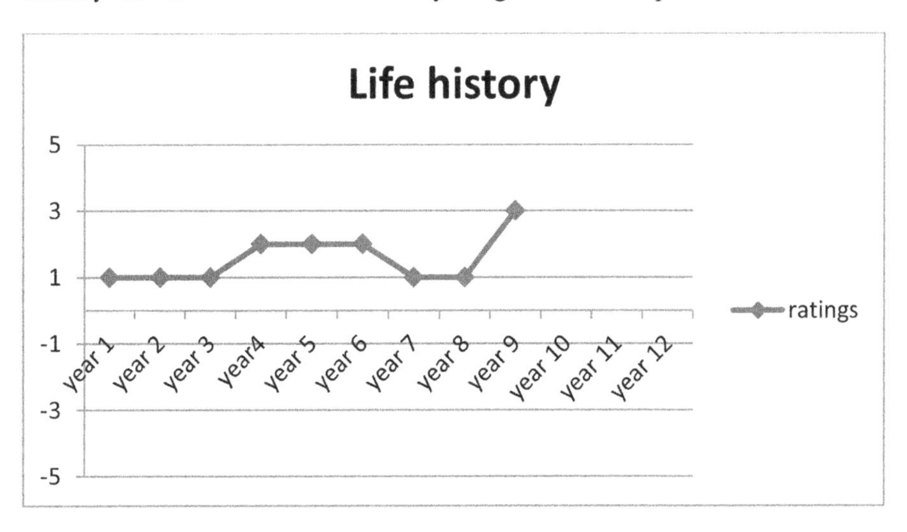

Figure 12. Liam's life history

However, on the TESI-ME, although he rates his overall stress as low (2), he scores high (6) on anxiety and somewhat high (4) on anger. His narrative reveals some significant evidence of negative affect. He appears to represent the classic profile of what Nardi and Steward termed 'quiet disaffection'. During the interview he was hesitant and thoughtful, and was not highly articulate about his own emotional landscape, often using 'I dunno' when asked to clarify his feelings. Nonetheless, this case demonstrates evidence of chronic but low level disaffection which is likely to impact the satisfaction and benefit he gets from mathematics, and possibly his willingness to engage with maths in a post-compulsory context. It is also a good case illustration of a dominant narrative.

Tentative Progress towards Competence

There is no evidence in Liam's life history of a dramatic decline into disaffection. Rather it bumps along just above the negative, gently rising and falling. This ebb and flow is well illustrated by the following:

> it was allright…I didn't do that good…I thought it was a bit hard…and…in year 4 and 5 and 6 I did pretty good and in y7 I thought it was a bit trickier… and then in y9 I find it a bit easier now.

This seems to be the pattern of Liam's progress, from "I didn't do that good", followed by some modest improvement, and then the process starts over again as the work gets more difficult. Like a number of others in this study, he appears to take time to get used to new ideas, but is confident when he does. When he/they then arrive at one of the 'cognitive watersheds' that so characterise progress in maths curriculum (sometimes algebra, but in this case fractions and decimals), he then gets into cognitive deficit ("I don't understand them", or "I don't get it") and this leads to a performance deficit, which he works over a course of years to overcome.

Here is another example:

> I didn't really understand that…but about year 3 …the last bit of year 3 I began to understand them…then up to year 6 I did good…then year 7 it got quite hard…and I'm quite good at it now.

This oscillation from positive to negative understanding is also matched in his affect in relation to mathematics. When asked if he likes maths, he replies:

> I did…er…but sometimes I didn't…because I thought it was a bit hard…

Or again:

> I quite like maths…I quite like it…I just don't like it when I don't get it…I like it quite a lot when I get it

When he doesn't get it, he describes how it feels:

> I panic quite a lot when I don't get it.

> it was just too hard…like fractions…I never like them…I hate them…I could do easy ones like quarters and halves…I just can't do like decimals

> …didn't like it…bit gutted because I wanted to do it myself…cos I didn't get it…

The process of recovery starts, although Liam is not firmly sure of how this happens:

> L – I started to get it in year 8…it's gone a bit easier

> Gl – what does getting it mean?

> L – about understanding it more…

Gl – and what helps you to understand it more?

L – the teachers and my mates…they help me…

Gl – how do they help you?

L – mmmm…I dunno…they just tell me a different way how to work it out (repeats)

There is one area, however, where Liam seems quite clear and articulate in relation to his disaffection, and that is in his explanation of his panic and anxiety. From line 52 to line 93 he describes a sequence multiple times (about 4/5 depending on interpretation). Since each of these has very little variation, it can be inferred that it is a strongly embedded pattern of experience.

The basic pattern goes like this:

I don't understand... I try... I don't get it...I ask for help...

Alongside this performative sequence is a parallel affective one. The negative emotions involved in this sequence are anxiety, panic or anger. He has a strong need to be able to do the set tasks himself (statement line, 52). But his keystone guiding motivational statement is:

I just wanna be where everyone else is

This appears to be his default motivational pathway, and he appears to be stuck in this pathway, since the majority of his narrative involves repetitions of this sequence. Being able to 'do it' appears to result in little pleasure in itself. The benefit appears to be that he is able to keep up with the others. At one stage he says "I'm always behind (the others)." Even if it is not literally true, it reveals how he often feels about the situation in his mathematics classes. If he doesn't 'get it' anxiety or panic ensues. Whether this is a direct result of not being able to perform the task, or as a result of the comparison with others is difficult to say.

I panic quite a lot when I don't get it.

By his own account, the sequence sometimes also ends in anger.

L – if I don't get it…if people are getting ahead…I'm always behind…and it gets quite annoying…I just wanna…be where everyone else is

Gl – why is it annoying?

L – cos everyone else can get on and I just don't understand…but some people get quite annoyed as well

The panic has turned to anger.

L – I get angry with myself because I can't do it…and then…I don't really… like angry…I'm trying to do it but I just don't get it…

> Gl – what do you do when you don't get it?
>
> L – I just ask miss…she just explains how to do it a different way…and when I've done it a different way then she explains how to do it the way that she does…so when I've done it the way that its easier I can do it the way that she's done it
>
> Gl – and does the anger go away then?
>
> L – yeh…then I start to get it and I do my work
>
> Gl – when you get angry…what's your first reaction…what do you do?
>
> L – I try and do it…for like a few minutes and then…if I can't do it I ask miss…if I still don't get it I get quite panicky cos I'm gonna get behind and everyone's gonna catch up with work and I can't.

This notion of comparative disaffection seems to be quite pervasive for many of the students in this study. For Liam, it is particularly acute. Perceiving oneself to be behind the others is a strong source of disaffection. In Liam's case it leads to anger – there is a sense of 'this is not right – it's not the way it's supposed to be.'

In motivational terms, his desire to be 'keeping up with the others' is an expression of self-mastery. To serve this need he has to be able to perform the tasks set as well (and as quickly) as the others, and so he approaches tasks in a serious-conforming frame of mind. When the goal is threatened, because he doesn't understand or can't do it, the arousal increases and he feels anxiety or panic. In his account, we also see that the sequence also sometimes ends in anger. This in turn suggests that not being able to do it disrupts his sense of conformity, and he reverses into rebelliousness, and this switches the emotion from anxiety to anger. However, since his account gives an undifferentiated description of these two emotions, it is not easy to be sure whether they operate in sequence, or whether the pathway sometimes ends in panic, or whether it sometimes ends in anger.

What is really interesting about Liam's account is that when the interview moves to a discussion of more positive affect (Lines 99 to 132), multiple replays of the same pathway appear, the only difference being that this time he can perform the tasks successfully, and so keep up with his peers, and the sequence ends in either relaxation or excitement.

Here is Liam talking about a sense of well-being – in this case 'relaxed.'

> – when I'm doing all the work…I'm doing good…and I'm with everyone on the work…and not behind…then I don't get panicky or angry.

But for Liam 'doing good' seems not to be a pleasure in its own right. The payoff appears to be keeping up with everyone else. Relaxation is just the absence of negative emotion caused by the disruption to his performance. In motivational terms, relaxation occurs in the serious-conforming state combination when arousal is low. And arousal is low because nothing frightening has happened.

But he always uses external measures of self-validation: his own performance compared to those of others. This external locus of control pervades his evaluation of his performance in any context:

> I do sometimes…sometimes I don't…in a test or something…I don't like them cos I know if I get a low score…then I know I've done bad…but If I get a high score then I know I've done good and I'm excited and relaxed…I know that I can do it.

Excitement is also sometimes involved in the sequence:

> I get quite excited when I'm doing the work…and I'm still with everyone and not behind…I get quite happy when I'm with everyone…if I'm the only one whose not doing it…then I get quite panicky…sometimes

It is basically the same script, but when he can do the work (and so still be 'with everyone') the affective pathway is positive. It is the comparison with others which brings the sense of pleasure: it doesn't appear to be the work itself. It also seemed to the interviewer that the description of relaxation was more or less identical to the description of excitement. When asked about the difference, Liam responded:

> – I dunno…not really…they're both the same really…I'm more relaxed when I get it and I'm with everyone…cos then I'm not behind and I know I can get on with my work…I'm more relaxed and I'm not panicky…cos when I'm panicky…I'm not relaxed and I can't do the work…and I'm always behind.

That description focussed on relaxed, so when asked to describe excitement, he replied:

> when I'm with everyone…when I don't need anyone to help me with the work…and I can just get on with it.

Again, it appears to be identical to the previous description. Yet in motivational terms, relaxation and excitement are quite different emotions, indicating quite different motivational states. What is happening? Relaxation occurs in the serious-conforming state combination when arousal is low. Excitement occurs in the playful-conforming state combination when arousal is high. There is no explicit evidence here to suggest that these two emotions occur in sequence, and the psychology suggests it is unlikely. What is more likely is that there are in fact two sequences, both of which happen at different times.

In the card sort procedure, Liam chose 'feeling part of the group' and 'feeling cared for'. When asked to explain the former, he again re-played the same sequence:

> when I've done all of the work…when I'm with everyone…and not behind… that's when I'm excited and relaxed.

When asked about not feeling part of the group, he says:

> no…I just feel …(PAUSE)… like I shouldn't be there…cos I know I'm not… like with the group…and I know I'm not doing good…then I can't concentrate and I get panicky.

Again, not being with the group is the controlling motivational issue. This is the external cue that he is not doing 'good' (sic).

When asked about feeling cared for, he says:

> feeling cared for…that's like when people are helping me…and I know… when I'm getting panicky…my teacher or my friends like say aahh come or let's do this… shows me another method of doing it…then I get relaxed…then when I've done all the work I get excited…most of the time it's friends cos… sometimes they can't help me when they're trying to help me a lot then…they can't explain to me how miss does it.

This may be a clue to the affective sequence.

Step 1 – Serious-conforming + can't do it, results in increased arousal resulting in emotion of panic

Step 2 – I get help, then I can do it, resulting in lowering of arousal, resulting in emotion of relaxation

Step 3 – I get on with the work, I reverse to playful state, being able to do it all increases the sense of arousal resulting in excitement

Although this sequence, and its variations, seems to dominate Liam's experience of mathematics, he appears to be motivated to try, and to put effort into developing the competence. Although the motivation to keep up with his peers (and not be seen to be behind) does dominate his experience, it is not the only aspect of his motivational landscape.

> cos when I grow up I'm gonna need maths…if I get a job…I can work out money…and stuff like that…cos if I don't carry on then I won't…cos if I'm in a shop or something…then I won't know how much money to give the customers…I need it…and English and French and stuff.

CHAPTER 13

THE LANDSCAPE OF DISAFFECTION

As can be seen from the case studies, the experience of disaffection can be quite individual, indeed, even idiosyncratic, so that the expression of it focusses on very different aspects of that affect. The experience of positive affect also plays a variety of roles in the experiences of these subjects. The factors that influence the disaffection also appear to differ widely, as does the expression of the experience. Individual patterns, or styles of response (like the dominant narratives here) are an interesting feature that will be commented on later.

Moving on from individual narratives, the focus will now shift to themes that are expressed across the experiences of the sample of interviewees.

That notwithstanding, in this chapter I will examine some of the common threads and themes that emerge from the evidence as a whole. Given the theoretical and methodological position of the study, one way that this can be usefully achieved is to use the Reversal Theory framework to capture the qualitative aspects of the subjective experience of disaffection in terms of motivation and emotion.

However, there are some aspects of the experience of these young people that seems to deserve attention in their own right.

DECLINE INTO DISAFFECTION

Take, for instance, the notion of the decline into disaffection. What seem to be the causes, or prompts that signify the onset of disaffection? Three themes emerge consistently throughout the data.

Adverse Comparison to the Performance of Others

The first of these is the adverse comparison to the performance of others.

Performance in mathematics seems to hold a significance for young people in a way that does not appear to apply to other school subjects. Pupils and students are acutely aware of their performance compared to others. Young people appear to make these comparisons unprompted, of their own volition. If the comparison is adverse, this leads to negative affect and negative self-evaluation. In this way, negative self-evaluations can lead to negative self-image, self-beliefs, and these in turn can lead to impaired involvement and impaired future performance, in a downward spiral.

It seems clear from this evidence that many pupils compare their performance (or more strictly, certain aspects of their performance such as speed of processing) to those of others. They are acutely aware of the adverse implications of such comparisons. These implications are about an important part of their identity, which is intelligence (being clever or 'dumb'). For many, personal reflection on adverse comparisons is enough. However, it can be made worse if other students react to this, and this seems to be an occasional aspect of student/pupil subculture.

> Yeh…because you…you kinda feel a bit down in front of them because they know more than you. (Pat)

So for Pat, presumably, there is a constant reminder that others are 'cleverer' than her.

> …I'm more relaxed when I get it and I'm with everyone…cos then I'm not behind and I know I can get on with my work… I'm more relaxed and I'm not panicky…cos when I'm panicky… I'm not relaxed and I can't do the work… and I'm always behind. (Liam)

So, keeping up (or appearing to keep up) is important. Keeping up is the condition for Liam to feel relaxed. If not, he panics and this impairs his performance. Not keeping up for many means feeling dumb. Being seen as dumb by the others leads to a sense of exclusion from the club or group. It also makes students vulnerable to taunts from those who are seen to be better.

> that's another thing…most of them are cleverer than me…and they used to put me down…they would like ring me up…they weren't really there to say…you know what…try harder…they just said you're dumb anyway…who cares…(Adnan)

The evidence in this study shows that when these adverse comparisons are systematised, in the context of the class, the consequent negative affect is amplified. Time and again in this evidence it can be seen that pupils feel labelled by their performance, and this becomes particularly acute and public when tests and exams are involved. This is not the same as test or exam anxiety (although it is real enough, it happens prior to or at the point of testing). It is more to do with the consequences of not doing well. Many of these accounts relate decline to the experience of SATs or other tests. The case of Anna shows the overwhelming effect that being labelled (in her case) C/D could have.

One way that pupils are clearly and publically labelled is when schools organise pupils into groups or sets based on ability.

> so…it's definitely one of the worst things I've ever had to go thro' in maths… is being put in ability groups…(Eve)

> I think that sets…having like sets in the classroom is not…I don't think it's fair on children…because then they feel that they're put into like whose good and whose bad at it…kinda thing. (Pat)

It seems clear, that for many, being placed in a lower set labels them in terms of intelligence or ability. Being thus labelled, it is easy for pupils to internalise this into a mathematical identity as 'being dumb'.

There is also evidence here that labelling exists in more subtle forms, and this involves less a public categorisation, but more the way the teacher behaves differentially to those who can do mathematics, and those who are perceived not to be able to do it.

> even tho' they wouldn't say to you...this is an ability group...you'd sort of work it out for yourself...easily...it's just...obvious isn't it...it's just completely obvious...(Eve)

Eve goes on to talk about:

> their frustration at having to explain it over and over again to you...and that's something where it knocks your confidence down... you sit there and you... uhh don't wanna ask...because you don't want to get frustrated that I can't get it...and that's how I always constantly felt that I couldn't ask because...of their frustration at having to explain it to me again...(Eve)

Even though this is Eve's perception, and may not be the intention of the teacher, the effect of the teacher's behaviour and approach is all too apparent. Although the hurt is heightened by being labelled so publically, as is the case with marks and grades, the evidence in these cases shows that students will look comparatively at the performance of others, and label themselves as 'failures' or 'dumb'. It is these strong implications for the sense of self and of identity that seem to be so damaging.

The Maths Changed – Got Harder or More 'Serious'

On the basis of this evidence, there appear to be certain 'thresholds' (Pampaka et al. 2012) when the mathematics that is taught changes in nature. It changes for the worse, in the sense that it becomes harder and hence less attainable. There are numerous accounts that dips in affect are influenced by changes in the nature of mathematics in the class. Algebra, fractions and decimals are topics that seem particularly to cause anxiety. New topics are introduced, and for some of these students, either they appear to make little sense, or they are unable to gain the traction of competence. The symbolic or abstract nature of the topics increases the cognitive load.

The Teaching and the Teacher Changed

As the mathematics itself changes, so the way that it is taught is also perceived to change, and this seems particularly acute at or towards the secondary phase. The changes are often subtle, involving a move to more individualised work. For many, it becomes more boring. Students perceive that they are left on their own to cope more, that there is less help, and the loss of working in groups with friends is lamented. The materials used in the classroom seem to reduce to exercises in books or 'just another worksheet'. The classroom and the process of learning mathematics seems somehow to become a less human enterprise.

because the teachers…they didn't explain properly innit…they didn't say how to work it out…they just give you worksheets…and you do it and they mark it…(Nina)

yeh…I can't…just sit there and write in a book…and there were times in year 4 to year 5 and we had to read…out of books and write…read and write…and that's all it was…I did not like that…I prefer more practical work…learning in a practical way…you understand more…you don't just look there and do the work…you actually get to experience. (Adnan)

In addition, there is an impression amongst many of these students that teachers pay less attention to those who 'don't get it', compared to those who do. They feel ignored and not cared for.

E – because I felt myself constantly being pushed down…when I felt I should have been pushed up…

Gl – how do you mean being pushed down?

E – like…I have noticed that a few of.my teachers would focus a lot more on the ones who were getting it rather than the ones who weren't, and I found that a bit strange…

Gl – so you were getting ignored?

E – yeh… which was really strange and that's why I felt like I just got dumped on somebody else…because the actual teacher didn't want to help and wanted to push the other students further …(Eve)

Alongside what is taught, and the way it is taught, there is evidence here that the personality of the teacher is brought to bear in the classroom in a qualitatively different way. This aspect is less easy to discern as a trend that increases through time, since many declines clearly relate to a change in teacher, and these can happen at any time. Losing a favoured teacher or gaining a teacher that the pupil doesn't get along with is often associated with decline.

Here are Helen and Masud:

you know like when you get teachers…and you get on better with them…and they teach you more…I think it depends on the teachers as well. (Helen)

but…because of the teachers innit?…like If you get a teacher which tells you do the work…like if you do something wrong he doesn't shout his head off… get angry…with some small minor things…like I had a teacher…every small thing you do…she'll just shout her head off like…she'll still move you… like…I dunno. (Masud)

Meena's description of a boring teacher is a good example of the way teachers' behaviour is perceived to change. It seems that encountering teachers who lack some of the qualities and habits that pupils enjoy seems more likely to happen in the secondary phase.

Contingencies

One interesting finding (but obvious in retrospect) is how often events and situations outside of the classroom effect pupils. In this evidence decline is related to (among other things) illness, moving home, parents leaving home, and other such disruptions.

As pointed out above, and as evidenced in a number of the subjects, it seems important to point out that disaffection is not a condition that is irrecoverable. There is evidence here in a number of accounts (based on the life history data) that students do achieve some level of accommodation with school mathematics. The experience of coming to college for a number of these students is the opportunity to experience a more motivationally positive and rich experience. According to this evidence this recovery from disaffection always seems to involve two factors. The first is the recovery of competence. A diet of mathematics too often characterised by failure and not being able to do it or understand it, is replaced by moderate success. Secondly, recovery seems always to be associated with a more positive relationship with the teacher and with enjoyment in the mathematics itself.

DISAFFECTION FROM A REVERSAL THEORY PERSPECTIVE

The experience of disaffection is not unitary. It exists and is felt in context, and is therefore holistic. In this way it crosses the boundaries of simple classification. The holistic and individual experience is reported in context in the case studies. In this section, it will be instructive to disaggregate aspects of that experience, and to present the evidence in terms of motivational and emotional themes. The Reversal Theory framework can be used as such a classification.

The Serious State

A great deal of the evidence of the experience of disaffection relates to the *serious* (telic) state. The serious state is about purpose or significance. It is about being motivated to achieve separable outcomes. In the serious state it is important to us that we feel we are making progress towards a valued outcome, or that we succeed at assigned tasks. If the likely or actual outcome is unpleasant or if we fail to make progress towards a favourable outcome, then a sense of danger ensues, arousal increases and negative affect follows. Since danger involves risk, and risk puts favourable outcomes in peril, people will be risk-averse in the serious state. Emotions in the serious state relate to the experience of arousal. Negative high-arousal emotions include fear (or anxiety, panic) when in combination with the conforming state, or anger when in combination with the rebellious state. Negative low-arousal emotions include boredom or sullenness. Positive emotions include relaxation (or relief), when arousal is low.

The evidence here suggests that students do need to understand the purpose or importance (i.e., felt significance) of what they are asked to do. Many students evaluate this in terms of utility. The goals that students adopt can be multiple and can exist at many levels, from individual task, to passing an exam, to career and life in general. Where a student hasn't adopted this sense for themselves, or where the teacher has not helped them to create it, a sense of meaningless or hopelessness follows.

Meaninglessness comes about because the student has no sense of significance attached to learning mathematics (or at least to a particular topic or task). Hopelessness is an expression of feeling unable to achieve the desired goal.

> if you know how to count…what's the use of doing big maths an that? (Helen)

> Gl – did you know you were not going to do…

> E – yeh…at that point…even tho I was ill…I sort of slacked more in it… because I thought even if I was well…I wouldn't pass it anyway. (Eve)

For many students, 'real life' applications of mathematics not only create interest, but also add to the utilitarian value.

> and it's like…hold on a minute…do you really need this in the future or is it more got to do with money and (?)…like percentages…like you need them when you go into a shop and there's 20% off…you need to know percentages… but fractions and stuff…do you really need to know this…like…I don't know. (Dena)

> mmmm…Maths at this stage is different cos it's more reality…cos what we're doing now is use of maths…is mainly I like real life situations…which is better cos I prefer real life situations…cos I know I'm not gonna learn something that I'm not gonna use in my everyday life…cos I'm not exactly gonna go to a shop and say 'give me one third of something'…(Adnan)

But if activities are evaluated only in this utilitarian way, they can be found wanting:

> yeh…one thing algebra…I'm sure you've heard of algebra before…a letter is a number…I mean…in a real life situation how often would you…ever realistically do that…as far as I know…I don't know…if I do find out it's a different story…cos I'm not going to go to a shop and say I have a amount of money…if I buy this which is £50…how much change would you give me? (Adnan)

Part of the problem with such explanations, is that they risk reducing 'viable' mathematics only to the very basics.

> you don't need most of the things in daily thing…do you know what I mean… like angles and stuff like that…how often do you measure angles in a day…

> ok…I might have to as a designer…like the pattern block and stuff…but other than that…like…a normal person…you probably just need it to count your money…you probably need it just to help you…(Anna)

But we are reminded that when students are able to make connections to some bigger purpose it makes the activities all the more motivationally rich for them. It should also be noted that there is ample evidence here that students can be sustained in the longer term to achieve a goal, even when the experience of doing it is unpleasant, if that goal is seen to be valuable enough in itself. That there are a large number (24) of references to 'purpose' in the data indicates the importance of this aspect of motivation to students.

There are two negative emotions associated with the serious state that are very prominent in this evidence. *Anxiety* occurs when in the serious-conforming state combination and felt arousal is high. It occurs when the outcome in the moment is negative or threatening. *Anger* occurs when in the serious-rebellious state combination and felt arousal is also high. There is a close relationship between the two emotions in that anxiety can be transformed into anger by a reversal between conforming to rebellious states. This pathway seems to be quite prevalent in the evidence in this study. First, the evidence on anxiety.

> I just remember every time it would be maths it's just like…anxiety would be horrible… because…it's…it's…it's embarrassing…you know if a teacher says 'can you tell me what this is?' …you're sat there completely blank even tho you're concentrating your hardest…and they ask you a question and you're like uuurr…and your whole class knows it and they're facing on you and you don't know it…you tend to feel really embarrassed…(Eve)

Eve's use of the word 'embarrassing' is as a catch-all term that is used by many students. The situation is embarrassing because she can't do the task (the humiliation associated with lack of competence). But it also becomes a sense of shame because her predicament is visible to all of her peers in the class. This experience is common.

> (on stress) I try to get the work done and everything but… it's stressful if you can't do something and everyone is ahead of you and you're still stuck on something like…(Harry)

Eve also talks about panic when encountering a task. It's not that she has tried and failed to get the answer that triggers the anxiety. She says "if it's something like algebra…I'll panic". This suggests that there is an evaluation based on self-mastery ('I won't be able to do it because it's algebra), and this in turn causes a shift of focus to the outcome (serious-conforming) and the anxiety follows.

Anxiety can also occur if a bigger sense of purpose is seen as in danger.

> because maths is important…I know it's something that you need in life…you need a good …like…maths grade…cos everywhere now involves maths… depending on which work you wanna do…it involves it so I know it's very

> important…so that's probably why I get really nervous cos I want to do well cos I know it's very important to have…(Meena)

The evidence here is consistent with the notion in the literature that anxiety inhibits performance (Hembree, 1990).

> there is one problem with me…whenever my exams come I stress out a lot…I just panicking a lot…what I'm gonna do and stuff…so…but yeh…I can do it…but I'm stressing too much that I can't do it and what I'm gonna do and etc…so. (Nell)

> oh…when I panic…the first time I panic…I look at the paper and I think how am I gonna do this paper…it looks so complicated and I only got one hour and I need to write so much stuff…that's the thing I first panicking. (Nell)

Interestingly, when Nell does get on with it, the panic subsides, and she starts to think maybe she can do it.

For some of these students, the anxiety carries on, and becomes more than an in-the-moment experience.

> apparently I've got anxiety my doctor says…I'm not too sure what it is…but panic…ignoring that…I do sometimes panic…when you get …exams…kind of freak out like shit…I don't think I'm gonna pass this…(Anna)

> me being stressful makes me worry a lot…like…it just makes me worry a lot… and when the exam comes up…maths and that…because I know it's important and I need to do well in this…I tend to not sleep…I tend to not eat. (Meena)

It seems that the key source of anger is being unable to 'do' the mathematics.

> Anger? I get frustrated. When I don't understand…I keep forgetting. (Dena)

> Anger – I'd say about 4 …you can get angry if you don't know something…if you can't understand it. (Harry)

> I get angry with myself because I can't do it…and then…I don't really…like angry…I'm trying to do it but I just don't get it…(Liam)

For these examples, the desire to make progress in the serious-conforming state is extremely strong, and lack of progress leads to frustration. The word frustration appears in many of these accounts of anger. I interpret frustration as the feeling associated with the denial of satisfaction in a state. Frustration is also a known cause of reversals, and a common sequence involves the building up of frustration (and arousal along with it), and this triggers a reversal from the conforming to rebellious state, with the associated emotion of anger. In this way, it can be seen that anger often arises as part of a sequence of negative experience.

> I think its sheer frustration…you know when you're doing a topic and…you somehow can't seem to grasp it…or you want to grasp it…like we did pi

> charts…the other week…and I wanted to so badly to get it…and I knew that I could get it…but my whole attention was just going all over…because I was focussing so much on trying to get it…rather than just actually getting it…and I think it's that frustration that really annoys me…when you're trying to do something so you can't get it…(Eve)

> well…like sometimes when we're doing like decimals…I can't do 'em… nothing…and actually anger…and even how much she explains it to me I can't do 'em because it just doesn't register…then I get frustrated and then I'm not doing the work…because obviously I can't do it (laughs) (Helen)

The sequence is also explained well by Alan:

> Gl – when you get stuck…does the panic come before the anger…or do they come together?
>
> A – I think the panic comes before then…the tension rise…
>
> Gl – and then the anger?
>
> A – yeh

For others there is some external referent that causes the reversal, and this is usually comparison to others. Masud gets angry when he can't do it (self-mastery, losing), and his feeling of humiliation is compounded when he sees that others can. This seems to cause a change of focus and a reversal to the rebellious state. The indignity of his incompetence is replaced by a sense of unfairness, which triggers the emotion of anger.

> because there's a reason behind it…how come I never learn how to do this and everyone in the class can know how to do it. (Masud)

Others experience a similar pattern.

> (angry) I try and do it…for like a few minutes and then…if I can't do it I ask miss…if I still don't get it I get quite panicky cos I'm gonna get behind and everyone's gonna catch up with work and I can't. (Liam)

> it's like…I can't explain it… like nervous…like can't do it…like you know when you can't do it…on the edge…cos everyone else can do it…your looking around…everyone's doing it…and I'm sitting there… (Helen)

What is interesting about much of this evidence on anger, is the control mechanisms that appear to be available to many of the students. All teachers (and parents) will be familiar with overt expressions of anger from young people which are often expressed in a dysfunctional way. But what characterises many of these accounts is just how much the anger remains internalised. It is often 'quiet disaffection', and strategies for dealing with the anger are many and varied.

H – normally I'll be quite angry if I can't do something…or like homework… I'll try and do it…it just gets a bit too much…and you just get angry

G – and what do you do then?

H – …I normally go outside and smash a football as hard as I can…normally helps but…my mums taught me a lot because I used to just like break things… she taught me whenever I'm angry just run up to the park and do what you need to do…if I get angry I just go up to the park and play football and it normally calms me down a bit…she helped me a lot. (Harry)

Liam's strategy is to ask Miss. When she explains it, he can do it and the anger subsides. Alan gets practical:

Gl – and how do you deal with the anger…when you're stuck…what do you do to get yourself out of it?

A – thanks to technology…you use the computer

Gl – and does that work?

A – yes. It does. Really helpful…you can go on Youtube and solve equations… it becomes a real teacher then

A strategy that may be more common is Helen's:

(laughs) I throw a strop and I can't do it so I don't do it…I'll sit there drawing

The feeling of anger is by definition unpleasant, so it is perhaps not surprising that young people develop strategies for dealing with it. For some, though, it sustains over longer periods, and becomes a more prevailing mood. Anna is a case in point.

not being able to get that B…that's what makes me angry…and I don't get why I can't get it…is it the teachers…is it me…I dunno…I understand the work…I just can't seem…

It also seems that in those cases, the anger can be used to spur activity to overcome the source of the anger (in Anna's case by gaining the pass at GCSE).

The Playful State

The paratelic or *playful* state is opposite to the serious state. The value of this state is enjoyment and having fun – doing things for their own sake, for the pleasure of the activity itself. In this state a person is likely to be spontaneous and creative. Since the attention is in the 'here and now', people are not mindful of consequences or outcomes in this state, and so risk (or what might otherwise be perceived as risk) is not present in the phenomenal field. In relation to the controlling variable of the somatic states, people will be arousal-seeking in the playful state. If arousal is low, this will be perceived as *boredom* (when combined with the conforming state) or *sullenness* (when combined with the rebellious state).

When students are in the playful state, and high arousal and excitement is unavailable to them, they will experience boredom. There is more evidence in this study of boredom than of any other negative feeling or emotion related to school mathematics. This suggests, as indicated above, that these young people spend a lot of time in the playful state, but are unable to find legitimate outlet or expression of it in the context of the activities of the mathematics classrooms. In some circumstances of low arousal, when combined with the rebellious state, students will experience sullenness, which is expressed perfectly in the modern idiom as 'whatever'. It is boredom with an overlay of negativism. Since students appear to be in the paratelic state so often, it is perhaps not surprising that students so often express this in inappropriate ways. Since high arousal is the need, students will create situations to increase arousal, and disruption, winding each other or teachers up, are very efficient ways of doing this. It seems that many teachers don't appear to understand that they could harness this playfulness into productive activity by introducing elements of curiosity, gaming, unfamiliarity, something out of the ordinary, to increase the arousal in a positive way, thus engaging the interest and engagement of the students. When they don't do this students get bored, and the two most cited reasons are 'lecturing' (listening to boring or bored teachers), and repetition.

> cos once you know it…you can't be bothered to keep doing it again and again…(laughs) (Pat)

> mmmm…probably…like I said before…I've been doing it for ages…since year 3 so I do get (??) bored…especially like doing that now…I do the same stuff as before…like in year 11…I mean like in year 10 GCSE. (Nadia)

> no… I just don't find it interesting…like…the way that the teachers explain it…it's like giving a lecture…and when someone gives a lecture…you tend not to listen…because it's just one tone of voice and you're just going on and on… and I just tend not to listen. (Meena)

> there's actually been like a science thing about it ain't there…where…about the voice…it actually helps them to learn…and it's actually true like…when you're sitting there and someone's got such a dull voice and you're listening to them over and over again…it's just…like…you switch off…because…erm… you don't feel like it's something to listen to…does that make sense? (Helen)

When students get bored, their need for distraction means that they react in various ways.

> I don't sit next to anyone just now…like you've seen me sat on my own…like there are plenty of seats…cos I feel like I get distracted…maths is boring so I'll get more distracted. (Anna)

Boredom? I tend to switch off a lot. I need a break. In fashion you're doing stuff, music…time flies. In maths…you're trying to listen to the teacher. (Dena)

E – I end up going back to being bored…because I'll just sit there…I'll just be like uuugh…because if I can't do it I'll just give up on myself…sometimes…

Gl – because it's better than being angry?

E – yeh (Eve)

It is a very fine line for teachers to get the balance between the need of students to achieve competence, and the need to avoid the repetition and drill which induces boredom. The hints in this data are that an unvarying diet of paper-based procedural exercises which replay the same skills are at the root of much boredom.

The Mastery State

In the *mastery* state, the operative value is power and control. Students need to be able to DO. In the context of a mathematics classroom, they need to feel that they can perform the tasks set successfully. When they can do this, the emotion is *pride*, and the feeling is confidence and being in control. In a competitive frame, being able to do better than others feels like winning – transactional gain. When they are unsuccessful, they feel powerless and helpless. The associated emotion is *humiliation.* It is also associated with the term embarrassment that is used so often by pupils and students.

it makes me feel like I'm dumb and stupid because of my grade…and I feel like…feel like…when people go…ah…what you doin at college and you go maths gcse or whatever…I'm ashamed to say I'm doing it because maths… is like very important…and I just feel dumb and stupid…and if someone else knows more or got a better grade. I just feel like ohmigod I don't want to say mine…(Meena)

sometimes when I don't understand something… feel a little…I feel dumb… I've got very low self-esteem…and this has been in the recent…a year or something…I've developed really low self-esteem…(Adnan)

like…humiliation sometimes…when you sit there like I say…you look back… and you think everyone else…and I can't do this an I should be able to do it (Helen)

because they think when I ask for help…other student maybe laugh at me… that I'm dumb and stuff…makes him or her feel bad and actually break the confidence. (Nell)

> that's where that ones come from...now if I don't learn something...I feel very... ashamed and embarrassed...they're just gonna call me dumb or something. (Adnan)

This feeling of humiliation can easily turn to hopelessness:

> because I feel...because I found every time I failed year 6, year 9...all the years...I failing...like...if I flop this year...what's the point waking up early every morning to learn and then every time you fail. (Masud)

It can be seen from this evidence that not being able to succeed at a task makes students feel 'dumb'. This is compounded by having a competitive or comparative outlook, which makes it feel like losing. Shame and embarrassment are terms used to describe the feeling of humiliation.

The Self-Sympathy State Combination

In the *self-sympathy* state the value is love or caring. In this state a person wants to feel that they are loveable, and that they are cared for. Since this is a transactional state, the values and feelings relate to human relationships. For a student to feel cared for, a teacher has to show a human and empathetic side. When this is not apparent, a person can feel unloved, not cared for. The emotion is *resentment*.

> Gl – how do you mean being pushed down?
>
> E – like...I have noticed that a few of.my teachers would focus a lot more on the ones who were getting it rather than the ones who weren't..., and I found that a bit strange...
>
> Gl – so you were getting ignored?
>
> E – yeh... which was really strange and that's why I felt like I just got dumped on somebody else...because the actual teacher didn't want to help and wanted to push the other students further ...(Eve)

This human aspect seems to be an important part of a positive and motivationally rich climate for the learning of mathematics. Good teachers do it naturally, and gain satisfaction from doing so. The empathetic and caring aspect of classroom climate was palpable, for instance, in College G. For other teachers, it may be that they just don't understand the motivational importance and significance for students.

POSITIVE EXPERIENCE OF MATHEMATICS

One of the most interesting findings in this study is that all of the students in the study, however apparently disaffected, recount positive experiences of school mathematics.

However negative their account of their relationship with school mathematics, it seems that their journey is punctuated with more positive experiences. This suggests that they are motivated in a positive and fulfilling way at times, but they also gain the satisfactions of the states. Again, we can examine these aspects of experience relating to different motivational states.

In the serious state the notion of significance or importance of an activity like the learning of mathematics comes to the fore. This in turn will be bound up with a student's view of the nature of mathematics. In the negative, or disaffected version, mathematics can be seen by students as a pointless undertaking with little to offer over and above perhaps receiving change in a shop. For these students it is concerned mainly with learning to perform pre-determined procedures. For other students, though, their appreciation of mathematics is more sophisticated.

> because result is just a pass or fail…more important is knowledge and information (Nadia)

> yes…it's very important…everywhere you go in the world…maths is very important…whether you an accountant…business…it relates to almost every subject…so you need mathematics. (Alan)

> Gl – you have to think a lot? Is that a good thing? Do you enjoy doing the thinking?

> A – yeh…when you think it's good… you can think a different way…different things…also when you think a lot in maths that does make other subjects easier as well

> Gl – tell me a bit more…

> A – you expand your brain. (Adnan)

A key value of the serious state is a sense of achievement. Being able to perform certain tasks, passing tests or exams, even though it seems so obvious to say it, is vital to learning, and to maintaining motivation and effort. Perhaps because it happens less often with these students, it is highly satisfying when it does happen.

> because when we did our functional skills exam I got quite a high score…I got 140…so if I get 110 on the other two then I can get a B or a C. (Pat)

> I like sense of achievement…I like to know I've achieved something today… if I leave the classroom and I think I've not really learned anything today…I wouldn't like maths in that way…I wanna leave that classroom and feel I've learnt something today…I like to feel that a lot of the times…(Adnan)

> yeh…I did the sense of achievement one…cos I paired it up with the sense of duty as well…cos I'm doing obviously what is expected of me to do…cos that's mel's lessons and she's got like a lesson plan…sense of achievement is like…after I've done it…cos mel could give us something really hard to do…

to see who can understand or get a rough method of how to do it…and that's my sense of achievement…if I know I've got it right…with my first go…or bits of it right…I feel happy …cos at least I got something right…(?) …the whole thing…(Raz)

Relaxation occurs when in the serious state, and the level of felt arousal is low. This tends to happen when one has achieved a positive outcome, or when making progress to that outcome.

probably say obviously relaxed…when I know I can do a topic. (Eve)

(relaxed) …when I'm doing all the work…I'm doing good…and I'm with everyone on the work…and not behind…then I don't get panicky or angry (Liam)

relaxed…when there's something I don't understand…and mel explains it… and then I get it how she works it out…and then I feel calm…like relaxed that she's told me and I know what to do now…and I can revise it at home. (Nina)

Adnan's explanation is interesting in that he uses music to manage his own motivational state, and to maintain the serious state in the face of potential distraction:

it's just a quiet environment…one of the things that really used to help me in school…which we weren't allowed to do but I used to do it…put headphones in my ears…and I used to listen to relaxation music…or just any sort of music…first of all it would keep me away from distraction from others…I know teachers say you shouldn't put your headphones and stuff…I never used to do it while teacher was talking…put my headphones in…do my work…I'm not being distracted by anyone else…it used to help me do the work…cos if I haven't got my headphones on I'll be doing my work…I'll be looking around…having a chat with someone…and then do my work again…whereas if I've got my headphones on…I wouldn't even know if anyone was talking to me…I'm working away. (Adnan)

In the playful state, high arousal will be perceived as *exciting*, and satisfaction in this state will be akin the state of 'flow' as described by Csikszentmihali (1996). The importance of this state in learning, and particularly in learning mathematics, has been consistently ignored or under-rated in the research literature, focussing as it does on outcomes, results and goals, which in turn trigger a serious state of mind. However, young people in this study consistently talk about the need for fun, and the value of enjoyment as a motivating factor in their engagement with school mathematics.

Interest occurs in many ways in mathematics classes. It can be triggered by curiosity, surprise, intrigue, or questions with unknown answers. In the context of a mathematics lesson, many things appear to be able to excite the playful state, and be perceived as 'fun'. New approaches, new materials, new ways of doing things, something about the associations of the topic, are all mentioned. Another way is to literally make a task into a game. There is evidence here that pupils will do this for themselves either to put themselves into the playful state, or to seek satisfaction in that state.

These things enable students to engage the playful state, and it puts them in a frame of mind to explore and discover. It is no surprise that fun, interest and enjoyment seem to go together as a package in the experience of these young people, and one that helps and encourages them to learn having fun and enjoyment.

> ...I always like to have fun while I'm doing stuff in maths classroom...even in any other classroom as well...always like to be happy...of course it you're not enjoying the course what's the point to doing it...you see what I mean... because you feel like scared...if you're not really enjoying then you don't know what really doing exactly...and you just get sad and doing stuff...it's totally blank. (Nell)

Here are a few descriptions of the ways that teachers can capture the interest of students:

> when I'm not just listening to a teacher going on and on...I'd rather have a teacher involved in some sort of humour...or not just doing maths but for example using maths in everyday life...so it would make you think...do you know...the next time I come across the situation I can do it like this...and that's where excitement comes in and curiosity...like it's a real life situation and they're making it exciting by giving you this real life situation...(Adnan)

> at some point I felt a bit of curiosity because...I wanted to know...because there was so much important stuff...like dealing with money obviously...and then...I knew that would help me at some point in my life...I knew...I'd get a job...or if I'd get in a career like dealing with maths or...so I did get curious... and some of the lessons were interesting as well... ...depending on the teacher in the classroom. (Nadia)

The nature of the classroom activities can also influence this:

> I like to draw graphs because I like to make them colourful...I like to colour in and stuff...especially bar charts...and right now since this new course started...they have many different graphs ...(Nell)

> because it (primary)was fun ...it was different ...but even tho it was like cutting up and sticking ...whatever you're doing ...colouring ...matching things up ... you're still ...I tend to learn more from that than if you're giving me a lecture ... because as you're doing it you're still learning thro what you're doing ...(Meena)

By framing activities as a game, the playful state can be encouraged:

> Miss B ...she taught it as a game...for different people she put it in different contexts...like for me it was always football...like angles...and stuff ...she

used football to help me understand it…but for different people it was different things…she knew a lot about us…what interested us…some people who liked to have questions…then work it out…but other people want to do it a different way…she taught us all a different way…how we liked. (Harry)

yeh…I like the competitivity…and I like practical stuff…let's just say…games and stuff…things get more into your head…like this one game we used to play…prime numbers…who can remember the most prime numbers wins… and teacher would give you all the prime numbers…and the one who can remember the most out of all of them would be the winner of the game…we used to play that at least once or twice a week…which really gets it into your head cos prime numbers are something that are sort of hard to remember… (Adnan)

A number of students made a connection between having fun and working in groups. It may be that the social interaction itself adds an element of interest, but it is also possible that working in groups tends to happen with a qualitatively different type of activity (perhaps more open-ended or exploratory), and that this also excites more paratelic interest.

The only time I've had fun is when I'm working in a group. (Dena)

yeh it's just nice to be part of the group …if were all doing the work together it makes it a bit more fun… than just doing it sitting there in silence…(Harry)

well, we do work in groups which is nice…because it's activity work…it's something interesting to do…I find it quite fun at times as well…because you're socialising…makes it more better…I think it makes it more better… like easy to understand…because you can share your ideas and your thoughts of how to do… how to work out a problem…and you can find the best solution for yourself. (Meena)

This package of ideas that includes fun, excitement, interest, enjoyment can all be considered aspects of the paratelic motivational state. The way that students talk about it, and the importance they attribute to it, makes it sound like a qualitatively different way to learn mathematics, and one that is highly valued by them. It is perhaps no coincidence that students' accounts of disaffection make it clear that paratelic activity and pleasure is usually unavailable in those classes that represent the most negative experiences for students.

Not surprisingly, positive affect is associated with mastery of mathematics. The positive emotion associated with self-mastery (winning) is pride, and there are 18 references to pride in this data. Other associated positive feelings include competence (28 references) and confidence (9 references).

Pat is a typical example in that she reports that she was good at maths at primary school, and found it easy, and therefore she enjoyed it.

> It's like for example...teacher gives us homework...and I've finished homework...teacher marks it and I get like 100% or full mark...I feel proud and very happy about it...yeh. (Alan)

> proud? I get proud when I do something good...for example I got a 9...I knew I was improving. (Dena)

> I probably wouldn't say it was that regular... I mean...I felt proud last week because I got the highest mark I've ever got on one of the exam papers...and so I felt proud about that...but then ... if I think about it...it's probably the first time I felt proud...this...whole term. (Eve)

Helping Others

One of the more surprising findings in this study is just how motivationally rich is the notion of helping others. In such circumstances, the active motivational state will be other-orientation. It is mentioned often in the narratives, and the card associated with this was one of the most chosen. And students appear to be quite thoughtful in the ways that they do this.

> E – I do like to help other people...but not...but obviously help them in a way that I would like to be helped...

> Gl – which is how?

> E – which is obviously starting from how I'd work something out...and not say...'it's this'... you need to do this...like actually show them...like write it down for them and show them...so they can see it...rather than me just being like you have to do this...and let them write it down...cos if they don't know how to do it they're not going to know how to write it down...(long pause). (Eve)

> I won't let them just copy my work...I'll explain it to them in a different way that...the teacher has, just to break it down...so they understand...and it does actually work...like one girl...the girl who's sitting next to me right now... she's the one who don't want to copy my work no more...she'll ask me to explain it rather than the teacher...I'll break it down...make it more simpler... which makes me feel happy...that also it's helping someone...it makes me feel happy (laughs). It also makes me feel quite important as well that she'd rather ask me for help and to explain it more...rather than the teacher. (Meena)

Meena's account shows that not only does she get pleasure from helping someone else, there is a collateral benefit for her – in this case, making her feel important. Others report different additional benefits:

> because if I help others...cos I learn more better that way if I explain it to others by helping them...so then I can go back and say I told this person this... this...and I can apply it to my way...(Nadia)

> N – I always like to help people if they don't really know about it…because I do finish my work quicker than others…so if they really struggle with it then I love to help them out…not showing them attitude that I finished early so I'm really clever student…so I can't help you out…I ain't got that attitude…I just always want to help them out
>
> Gl – how's it make you feel when you can help them?
>
> N – I feel proud…I feel happy that I actually helped someone…and I actually helped that person who feel like good …and the person feel like ok…she helped me and I know now exactly what I need to do…so thats make me happy. (Nell)

> I also like helping others…if I understand something…someone else doesn't…I just help them out…it also helps me understand it better…that's why I like doing that a lot…also…as well as having fun and enjoyment…(Adnan)

> because apparently if you teach someone you understand it more yourself… cos you're repeating it…cos maths is about practice…you're technically just repeating it…it does help…if you help other people you learn yourself… (Anna)

Helping others is clearly a motivationally rich activity. By motivationally rich is meant that the activity offers the possibility of satisfying a number of motivational needs. Thus, one can be in different motivational states and be satisfied by helping people in different ways. By helping others, students can gain a number of satisfactions simultaneously.

Doing It My Own Way

The motivational state of rebelliousness (negativism) appears, not surprisingly, on the more negative side of the balance sheet in terms of the associated emotions of anger and sullenness. Expressions of negativism by students are also all too common in classrooms, and can cause behaviour problems that are difficult to handle. So it is good to report that positive and functional expressions of the rebellious state do have an important part to play in mathematics classrooms, and in terms of students' positive experiences of mathematics. The negativistic state is opposite to conformity, and can be seen as the need for freedom, for autonomy and self-expression. It is an important aspect of the development of the individual that they can define and assert aspects of their identity in contradistinction to the expectations of the norms prevailing in family and other aspects of their socio-cultural context.

In this study, evidence of this is found in the need for students to develop and use their own methods in solving problems.

> A – I just wanna find my own method and I feel freedom…doing it my own way

> Gl – sounds like the method is really important
>
> A – yeh…that's what maths is all about…method…if you can understand the method, then you can do the rest. (Alan)
>
> Let's say I've learnt something…and I figured out my own way to do it…a technique or something…anything…randomly looking at patterns and stuff… and figuring out things my own way…I wouldn't keep it to myself…I'm not that type of guy…I'll literally put my hands up and I'll scream out to the teacher…I've figured another way to do it…I'll tell the whole class…I wouldn't keep it to myself…(Adnan)
>
> I do tend to get competitive or like…tell them that answer is wrong…and say it my way. (Nadia)
>
> every student…every person has a different way of calculating…dividing, times, or like any percentages…like some people do it a different way… mostly the main way to do it…everyone does it the main way you're supposed to…which I do…and sometimes I change the way I try to work it out (Nina)

OTHER ASPECTS OF DISAFFECTION

Looking at the evidence of the experience of disaffection in the qualitative analyses, a number of interesting points emerge.

Volatility

It is useful here to adopt the distinction made by De Bellis and Goldin (2006) between global affect and local affect. Global affect is seen to be those aspects of affect that have crystallised into more hardened and stable positions. Global affect will encompass all aspects of affect, including disposition (like/dislike), self-efficacy (can/cannot do), as well as prevailing emotions which themselves crystallise into moods (e.g., boredom). These in turn will effect behaviour, and in the case of disaffection will include avoidance, feelings of hopelessness and helplessness, and emotions such as anxiety, humiliation and anger. For disaffected students then, global affect will be predominantly negative. Local affect, on the other hand, involves the more in-the-moment aspects of feeling and emotion. These more volatile aspects of affect are less recognised in the literature, and less studied.

Perhaps one of the most surprising and interesting outcomes from this study is just how volatile the nature of these students' relationship to school mathematics. This volatile nature seems to operate at a number of time frames. This fractal-like nature is an interesting phenomenon. One important finding of this study is that even strongly disaffected students can experience positive affect on a local scale.

In this study, sequences of local affect have been examined in detail, and two patterns of particular interest are commented on below. The work of Hannula (2006) and Op't Eynde et al. (2006) has been extended in that a number of such episodes have been reported, and the sequences of emotions that they represent have been given explanatory interpretations, using the Reversal Theory framework to understand in some detail how these sequences relate to activities and outcomes in the mathematics classroom, and how they relate to the advancement or inhibition of learning.

On a more general scale, the evidence of the life histories shows that more global affect also ebbs and flows on the year-to-year basis. A number of factors have been reported that account for these ebbs and flows, but the evidence in this study suggests that perhaps the most influential factor is the microculture within the individual classroom. These ebbs and flows have not previously been accounted for to date in research, since affective constructs such as attitudes and beliefs have been held to be relatively stable.

It is tempting to interpret the rises in perceived affect as a sort of 'recovery' from the condition of disaffection. Unfortunately, such an interpretation would not be justified by the evidence. Firstly, the rating seems never to rise to +5. Rises in affect are apparent, but they do not appear to represent a quantum shift to outright affection for mathematics. Perhaps a more measured interpretation would be the notion of respite rather than dramatic change. In these periods, students describe their experience of mathematics as better, or not so frightening. Feelings of competence and satisfaction are apparent, and the 'telic horror' of the dips in affect are to some degree absent or mitigated.

It is difficult to see any reason in principle why, given an accumulation of positive and affirming experience, a person could not make the journey from disaffection to affection with mathematics. Unfortunately, there is no evidence in this study that any student has made such a journey. What we see instead is that mathematics becomes tolerable for some. The recovered condition is usually predicated on a utilitarian accommodation.

The disaffection is deep-rooted, encompassing as it does attitudes and beliefs (including beliefs about self), as well as feelings and emotions. These entrenched aspects of affect have hardened into the identity, expectations and behavioural habits of these students, thus making them strongly resistant to change.

Emotional Range

Within the qualitative interview data, perhaps the evidence relating to emotions is amongst the most compelling. A wide range of emotions has been evidenced, and distinctions such as high versus low arousal; pleasant versus unpleasant hedonic tone; somatic versus transactional emotions, have introduced new theoretical ideas to both describe and explain such emotions. This study has demonstrated how these negative emotions weave in and out of the experience of disaffected students, and the disabling effects that they can have on learning.

In particular, the study has shown how anger operates in the lives of these students. Evidence has been brought forward to show the genesis of anger in sequences of participation in mathematical tasks. It is perhaps a new and interesting finding that much of this anger is kept internalised by students, and how good many students are at managing and regulating their own anger. This advances our understanding of such issues, and offers a counter to the generally held notion that low performers do not manage their emotions as well as higher performers.

This study has widened the scope of examination of reported emotions, also their relation to motivation and theoretical foundation of both, including the ways that they inhibit or enhance learning. In particular, low-arousal emotions such as relief/relaxation and boredom have been described, and these have been poorly reported in the literature to date.

Self-comparison and performance anxiety. There is strong evidence here that self-comparison to the performance of others is extremely damaging for those for whom the comparison is adverse. This is consistent with the evidence of Dweck and others. The evidence here suggest that there exist in British mathematics education, a number of ways, both formal and informal, in which such comparisons are a structural feature of our pedagogy, and are damaging for being so.

Effort and Other Metacognitive and Regulatory Skills

The role of effort and other aspects of metacognitive and self-regulatory skills are abundantly present in these accounts of disaffection. This evidence is perhaps a counter to the notion that low-attaining students do not have such skills to the degree that their more high attaining peers have. The evidence shows that not only do these students espouse effort, they also show evidence of the application of effort, even when it is very difficult or painful for them to do so. The valuing of effort is also shown in their attribution for success and (lack of effort) in attribution for their own failure. Many of these students view that success comes as a result of effort and determination.

As has been pointed out in the review of literature, students who fail to succeed at learning mathematics (and that includes most of the students in this study), are deemed less likely to have developed metacognitive and their related self-regulation skills. The evidence here suggests that on the contrary, many of these students have these skills, and sometimes to quite a high degree.

First of all, in terms of attribution for success, when asked what makes other students good or 'clever' at mathematics, these students suggest that it is hard work and effort. It is certainly not the case that most of these students (as predicted by Dweck and colleagues, for instance) have an entity view of intelligence. Here is Meena and Anna:

> no not really…everyone's born the same…I suppose it's the environment… …I think it's the environment and the atmosphere…that makes you clever. (Meena)

> I used to think that people are just born smart…but then I kinda got my head round it and now I just think…(HESITATES) …it's time and effort…you're not just born smart…its time and effort… it's how much you put into it… cos anyone can get good grades…you can…like if you went home and if you went through everything everyday…and you were that organised…you can get good grades…I think it's time and effort…but it's something I find really hard…cos im really disorganised. (Anna)

That attitude carries over into blaming self when things have not gone well.

> it takes a lot of effort…and probably time as well…probably I would say… with the group I was in…I was with people I knew, so I used to get a lot of distraction…I'm on about from year 9…I used to get a lot of distraction with maths…cos I used to be put in with my best mates an that…so we always used to talk an that…so thats why it put me down a lot…and other people used to achieve a higher…so I blame myself. (Nadia)

This respect for time and effort seems to play out in their approach to mathematics. Considering how painful and difficult mathematics can be for these students, it is humbling to note how they are still willing to try so hard to gain the success they wish for.

> after two years of struggle…you get used to it…(nothing else changes)…then you just try hard, hard hard…then you get used to it. (Alan)

> I mean… the teaching I had last year wasn't as good…but I got to the point where I needed to focus…I'd sort of come to the realisation that I need to do this for myself because I'm not going to get any help. (Eve)

> I just want to prove myself…I don't want to give up…and know that I've tried and not give up…even if it was…some other person in my position would probably just give up…but I want to prove to myself that I can achieve a higher grade, yeh (Nadia)

This effort does lead to success for some.

> for example…like equations…Some equations…you'll never get it…it's so hard…but some equations…it's hard but you know you get it in the end…by the knowledge…I've learnt how to do it…that's the reason behind it…will take you to the answer…I'll be positive like you've done it by yourself. (Masud)

When asked if they would study mathematics after college if it meant advancement in their job, here are a few replies:

> yeh…but I would do it…I would probably put everything I had into maths… you can…it's just time and effort…I'm not too sure…it's just a theory but…

> it's like business.isn't it...business is time and effort and that's how you get a lot of money. (Anna)
>
> I would try it...I *would* try it...I would take it...if it was an upgrade...(Helen)

Balance and the Importance of Positive Experiences

A good deal of evidence has been brought forward to show that life in a mathematics classroom is not always unremitting gloom, even for students who are disaffected. The evidence derived from the card sort shows strongly that almost all students experience some positive affect from time to time. Being able to 'do' (by which we take to mean, to perform successfully) the tasks set is a particular source of positive affect. When students can do the tasks set they report a sense of relaxation (an emotion related to achievement in the serious-conforming state combination, when arousal is low). It is also associated with pride (self-mastery, winning), and this seems to be highly motivationally satisfying when it is available.

A good deal of the evidence here relates to those (apparently rare) opportunities to be engaged in mathematical activity in a paratelic state. Sequences in which this happens are examined in more detail below, but it is important to note that opportunities for paratelic enjoyment are often at the centre of focus of positive experiences.

It has been noted throughout this study, the satisfaction that can arise from the positive experience of being able to help others. Another surprising source of satisfaction that has been reported here relates to the assertion of individuality and freedom through developing and choosing one's own methods.

It is important to note, however, that positive experiences – either in the moment (local affect) or on a more global scale (such as the rises in the life history chart), do not necessarily represent a change in the basic disposition of lacking affection for mathematics. For most, it appears to have more the characteristic of a temporary reprieve.

Lack of Significance

Perhaps a more worrying aspect of the motivational and emotional landscape of disaffected students encountered in this study is the felt lack of importance or significance of the mathematics they encounter. If students are not helped to be convinced that mathematics is valuable and useful, it is hardly surprising that they fail to embrace it more positively. The lesson for teaching practice is that mathematics must be presented as more than a set of tasks and procedures to master. Teachers need to place mathematics in its proper utilitarian, but also cultural and social context to make it more meaningful to students. The evidence here suggests that in many mathematics classrooms, motivation is demanded by a requirement for conformity or duty, rather than a case for felt significance. The challenge for teachers is to make this aspect of school mathematics more alive and relevant for students.

Importance of Understanding to Doing

Time and again in this evidence there is mention of the notion of understanding, and its importance to learning for so many of these students. Statements take various forms:

> If I can understand it, I can do it…
>
> If I understand it I enjoy it more…

It is tempting to interpret this as a natural preference for a connectionist approach to the learning of mathematics, and the evidence would support this to some degree. On the other hand, it is clear that what is often meant is that the student just needs to be given a simple and clear explanation of what to do. In either case, the blind application of procedure or algorithms is clearly not enough for most students.

Cognitive mastery ('I understand it') seems to be intimately related to performative mastery ('I can do it') in the accounts of these young people.

NOTE

[1] The notion of recovery or re-affection is discussed more fully in Chapter 13.

CHAPTER 14

SOME THEORETICAL PERSPECTIVES ON THE FINDINGS

There follows a number of themes which seek to move on from reporting data from the study to seeking to develop some theoretical perspectives on motivation and disaffection with mathematics. Four of these merit mention as a group. The four are: the development of cognitively-mediated structures; the landscape of disaffection; motivational pathways; and motivational climate for learning mathematics. What they have in common is that they are proposing structures that are derived from patterns in the data, or are informed by applying ideas from theory to the data, or a combination of both. At this stage, they are presented as propositions or conjectures. Whilst they are true to the data in the study, it cannot be claimed at this stage that they are universally true. Nonetheless, it is hoped that they have resonance and utility for other researchers in the field, and they may merit further research.

THE DEVELOPMENT OF COGNITIVELY-MEDIATED STRUCTURES

In the previous chapter mention was made of the influence of strong emotions, and the relation of such emotions to other more cognitively-mediated and more stable affective dispositions such as attitudes and beliefs. It is now possible to be more precise about the development of such structures, using current knowledge and by applying additionally a Reversal Theory perspective. McLeod (1992) mentioned the importance of the repetition of experiences in the development of such dispositions. What the evidence in this study shows is that they are learnt and developed within the mindset of specific motivational states, as shown below:

It is proposed that if the cognitive evaluation or comparison between perception of experience (feeling or emotion) and current attitude or belief are in alignment, then this loop will be a reinforcing or amplifying one. In that case, the attitude or belief hardens. If, however, there is an evaluative discrepancy between the two, this may cause a re-evaluation, and so cause the attitude or belief to weaken.

Evidence confirms that attitudes and beliefs expressed by students are often specific to a particular motivational state and exist within the frame of that state. Examples abound. Meena's belief that mathematics is best learnt when it is fun is related to her own need for paratelic enjoyment in the playful state. Liam's attitude to mathematics tasks as a competitive performance likely comes from his

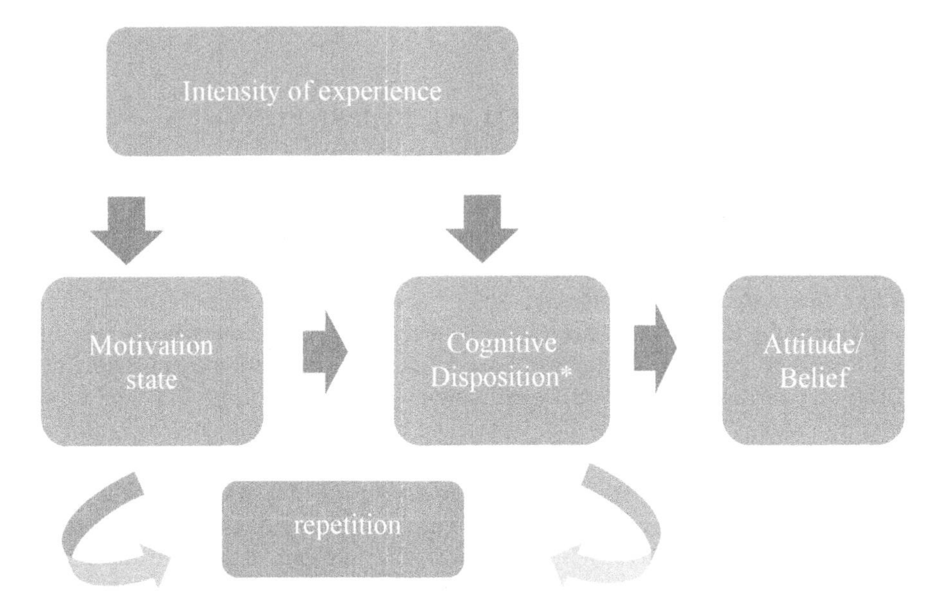

Figure 13. The evolution of attitudes and beliefs
(Cognitive disposition represents the existing structure of attitudes/beliefs and influence of socio-cultural context)*

strong (but often unsatisfied) need for control in the self-mastery state. There is evidence from a number of students that they feel 'teachers don't care for me'. Such attitudes and beliefs are clearly grounded in the students' sense of self-mastery (losing).

THE LANDSCAPE OF DISAFFECTION

To assist in this discussion it is useful to draw a map of the landscape of disaffection, even though it is necessary to bear in mind that this can be considered a simple (but not simplistic) representation of the space.

In the zone of disaffection the disaffection itself is consistently held and prevailing, not temporary. It means giving up; I do not love or like it; I will avoid it if I can. There is a grammar of systematic dislike, which is decisive and knows itself. The narrative is 'I will not, I am not…' It is an affirmation of the negative. Rejection is a behavioural litmus test of disaffection. This leads us to a fairly decisive description of the condition of disaffection.

In the zone of affection, the opposite conditions prevail. A person will consider themself a mathematician, or a learner of mathematics, and they will experience anticipation and enjoyment from learning mathematics. As pointed out above, those with this disposition seem to be rare, although this judgement may be influenced by the selection of the sample, and by the methods used. Nonetheless,

Table 13. The landscape of disaffection

	Disaffection	*Accommodation*	*Affection*
Overarching disposition	Negative – rejection	Volatile affect	Positive – attraction
emotions	Anxiety, hopelessness, helplessness, shame	Neutral (uncaring) punctuated by occasional positive and negative episodes	Excitement, pride, playfulness
Behaviour/ competence	Feeling 'dumb' Avoidance	Tentative mastery	Confidence, curiosity
motivation	Mainly telic	Utilitarian accommodation	Includes positive paratelic experience

– attraction +

it is reasonable to suggest that it is only the small minority who hold mathematics in real affection.

As indicated in Table 13, there appears to be a transitional zone where strongly prevailing disaffection is punctuated by more positive experiences. Although in the case of many of the students in this study for whom these more positive experiences are merely a reprieve, and do not fundamentally alter the underlying disposition, it should not be discounted that it is possible to move into such a transitional space where the underlying disposition is loosened.[1] Inasmuch as students here might be considered to be in this territory, it is clear that any positive affect is both transient and predicated on a utilitarian accommodation as described in the evidence.

The evidence from Lewis and Forsythe (2012) suggests that students can make the journey from the zone of accommodation to the affective zone. On the basis of the evidence here, no student in this study has made the journey from disaffection to affection. None of the students interviewed would voluntarily choose to do mathematics again. In other words, although they will engage with mathematics, there is little evidence that they enjoy or embrace mathematics. They will apply discretionary effort. However, in almost all cases reported, their fundamental underlying disposition is still disaffection. They are not 'cured'.

Other evidence suggests that it is possible to make the journey across these spaces. Watson and de Geest (2005) report pupils in their study become more willing and more enthusiastic in their approach to mathematics. Further, their accounts of pupils embracing challenges and difficult tasks, and accepting a level of risk suggest that it is possible to do so. Those results have to be treated with some caution. Evidence of willingness and enthusiasm is positive and heartening, but more detailed evidence around the affective landscape of these pupils would be needed to be sure that the transformation was as suspected, and more than temporary. There is some similar evidence in Boaler (2000). She reports the results of a summer school experience for students in the US. She reports some vivid cases who become enamoured with mathematics through this process, and a tangible improvement in their achievement scores on return to regular education. However, these gains are short lived. It is possible to interpret this data in a number of ways. Firstly, enduring recovery from deep disaffection seems difficult to achieve. Secondly, the influence of the microculture of the class, bringing together as it does all of the influences of curriculum and pedagogy and lack of genuine motivational connection through enriching activities, seems all too easy to reassert itself.[2]

It does seem, on the basis of the evidence here that the conditions for re-affection will involve the development of competence borne of understanding and pleasure in the act of doing mathematics.

STRUCTURAL INFLUENCES ON DISAFFECTION

In the model of the affect space (Chapter 3) the influence of the socio-cultural context is accounted for. This category is in itself a complex space, and multiple and complex factors can be identified which in turn will influence the affect space of any individual. It is not the intention here to give a full or comprehensive account of all of these potential factors, and how they interact to influence the experience of disaffection. However, three factors seem to have a particular relevance and importance, and will be discussed here.

Coercion

Coercion is one such factor. The notion of coercion is complex in this context. At the simple level, pupils in school have to study mathematics. The students in the Colleges are also conscripts and not volunteers. Students' scope for action is limited and they do not have the power of refusal. Not surprisingly, this immediately colours the motivational climate, in that mathematics has to be endured. Acceptance then becomes the key. Given the choice, and the way disaffection has been characterised here, it seems obvious that many (perhaps most?) of the students would not study mathematics by choice. But there are also more subtle influences at play here. It is not just what I am required to do, but why, that also influences the situation. The enterprise I am required to undertake is not just to learn mathematics to be a competent mathematician, (although it may be presented in that way within the classroom culture), but to pass an examination. The qualification becomes the sine qua non of the activity. All of the alternative functional, personal or cultural rationales for studying mathematics are subjugated to this requirement. This requirement of the 'other' not only distances it from my ability to choose it for my own ends, it makes it more difficult for an individual to develop healthy internalised motivation for doing it.

Duty

Secondly, alongside, but related to coercion, is the notion of duty. It has been remarked that a feature of these students' approach to studying mathematics is the degree of conformity which they adopt, given the analysis above. On the whole, it seems, these students want (for much, but not for all of the time) to fit in, and to do what is expected of them. What is expected of them derives in good measure from the socio-cultural influences around them. These influences are in turn both multiple and complex, involving generational, parental, and institutional aspects. In this evidence this is expressed as obligation, duty, although at times it exists without being articulated. Parental expectations are often articulated influences here, as is the sentiment of not wanting to let the teacher down.

Together, the two influences of duty and coercion work to create an environment where many young people feel that they have little voice and little choice. This can lead to feelings of being trapped into an undertaking which is not only unpleasant, but in which they have little or no chance of winning.

Instrumentality

The third influence which should be mentioned in this context is that of instrumentality. Instrumentality is the recourse to notions of separable outcomes by virtue of the utility or exchange value of achievements in school mathematics. Thus we see that students are convinced (sometimes) that:

- Being good at mathematics is a necessary skill for everyday life.
- Passing exams is necessary to open opportunities for further and higher education (including in other subjects).
- Mathematics qualifications are advantageous both in the acquiring of jobs and being successful at work.

At one level, all of these are valid reasons to study mathematics. But there is also a price to pay for an over-reliance on calls to utility as a motivator to learn mathematics, and the consequences can be insidious. So, because I need to do it does not mean I will love it.

Coercion, instrumentality, duty, represent the boundary conditions of the socio-cultural context within which disaffection occurs. Yet instrumentality does not produce positive motivation and affection. Although students may do mathematics, and they may even achieve a level of success, they are still, in a real sense, lost to mathematics. If disaffection remains, even within the narrow context of exam 'success', the limitation on the development of capability also remains. This is the insidious aspect. Disaffection always means a loss of hope, future, possibility: in short it is a loss of capability. It is an impairment, a disablement, an unrealised potential.

MOTIVATION

As Hannula (2012) noted: "Almost all work on attitudes misses important distinctions regarding its quality, simply focussing on the direction and magnitude of attitude" (p. 141). Much the same could be said of research in motivation in general.

It has been argued here that motivation is central to an understanding of the notion of affect, and hence disaffection. A theory new to the field has been introduced, and this leads us to challenge some of the current thinking about motivation. Motivation is described in the literature as 'an inclination…' (Hannula), or 'a willingness…' (Anderman & Wolters). Ryan and Deci (2000) talk about being 'moved to do something.' It is interesting to note that the metaphor is mechanical – motivation is a kind of force driving us to do or not do some task. There are a number of problems with these definitions. Firstly, 'inclinations' or indeed 'willingness', sound very like 'dispositions', and these suggest the notion of motivation is almost indistinguishable from attitudes in many definitions. Although we may be willing or inclined, it presupposes the question as to what are we inclined to, and why?

A further problem is the way that motivation is attributed to young people. Thus we hear that pupils are not motivated, suggesting that motivation is a binary condition that we either have or don't have. A variation on this is to talk about 'low' motivation. The problem with this is that it changes the metaphor – now we have a bucket that is at some stage of fullness or emptiness. This is not to say that intensity of motivation is not important. Our everyday experience tells us that sometimes we feel much more urgent about aspects of our phenomenal experience than at others. It is perhaps important to acknowledge that there is an everyday or common-sense use of the term motivation which implies positive attraction to some idea or activity, together with a degree of intensity. In this book, the more technical characterisation of motivation provided by Reversal Theory is used throughout. However, the common-sense notion will be characterised here as motivational connection or attraction, when considered in relation to a specific task or object of concern within a motivational state. The contention in this study, and verified by the results, is that motivation is not some amount of 'stuff' that we do or do not have. A much more interesting question than 'how much are you motivated?' is 'how are you motivated?' The evidence in this study shows that it is important to pay attention to the qualitative nature of motivation, rather than to be stuck on the notion of its quantification. A recognition and understanding of the different motivational states is one way to do this.

To give an example, a student who is bored, and displays signs of this boredom in a classroom, may well be viewed by a teacher as lacking in motivation. However, I would contend that it does not mean that the student is unmotivated, or that they don't want to learn mathematics, it is just that their motivational need at the time (for stimulus, excitement, etc.) is not being met in that specific classroom context. There is no lack of motivation, only of motivational connection to the task in hand. There are many tactics that the teacher could employ to meet the need, and provide the necessary motivational connection.

The notion of disaffection seems for many teachers to be bound up with whether a student can do mathematics. 'Can't do' is inseparably bound up with the notion of 'won't do'. Teachers make judgements on the basis of perceived attitude and behaviour. I asked teachers at school K to nominate the student in their class who was most motivated and the student who was least motivated, and to describe why. Here is the data:

'Least motivated, disengaged or disaffected…'–

Attempts as little work as possible

Dislikes reading and writing

J rarely brings appropriate equipment to the lessons

Appear not to see the benefit which results from being able to do maths

No focus

A finds it difficult to engage with the more difficult work

Doesn't like school and would prefer cigarettes

Comes poorly equipped to lessons and looks to avoid work where possible

Poor attitude. Easily distracted…will do as little work as possible although she is a good mathematician

'Most motivated…' –

Always does the extension activity

Attentive well behaved

Keen to try the most difficult work

Gives the impression that she wants to learn and improve

Focussed attention

Directs herself to parts of the task where she feels weakest

Tries to get all the work done

These few examples are instructive. They are a mix of behavioural and attitudinal, but what comes across is that the unmotivated students do not appear as willing. Discretionary effort seems to be the main yardstick by which teachers evaluate disaffection. Motivation, or lack of it, is for teachers, about visible effort. When students don't appear to conform to the requirement to show enthusiasm and visibly put effort in, this is described as a lack of motivation and not liking mathematics by teachers.

The evidence here shows that students, on the other hand, are often desperate to get motivationally connected with mathematics, even though they are disaffected and even though it is often painful, but the pedagogical offer often does not enable them to do this. This is a failure of teaching.

INDIVIDUALITY AND MOTIVATIONAL STYLE

The evidence of the case studies shows how all of the affective dimensions, that are so often studied in isolation, operate holistically and in a very human and individual way. The case studies demonstrate that there is not only one way to be disaffected. Perhaps it is a truism that every unhappy man is unhappy in his own way.

In the review of literature, the notion of trait was examined, and a case was made that it was too simplistic and inadequate to explain patterns in behaviour. Nonetheless, this is not to say that patterns over time do not exist. One of the primary weaknesses of the notion of trait, is that it makes recourse to habitual behaviour. The research reported here shows that student's engagement or connection (or not) with school mathematics cannot be classified into simple categories, and that responses and experience are extremely sensitive to particular conditions. Moreover, students can behave in very different ways, even in the same or similar situations, depending on the motivational state they are in.

However, there is here evidence of pattern and stability in student's responses, even though they cannot be characterised by the notion of trait. Such individual patterns can be labelled as motivational style.

In Reversal Theory one of the key patterns of individual motivational style is called dominance. It is based on a psychometric measure of frequency of reported occurrence of being in each state. The difference between the two scale scores for each state in a pair is standardised, and the result is called dominance. It is thus a measure of our preference for one state over another. The non-dominant state is still active (for most psychologically healthy people). Dominance has been studied extensively in the literature, and is known to predict certain behaviours (Apter, 2001). Interesting though the notion of dominance is, no data has been collected in the current study, although it is probable that dominance may have some interesting properties in relation to the ways that students engage with learning, but must, for the moment remain a question for future study.

Within the data, some grounded patterns have emerged and are evident in the case studies. One such is the notion of dominant narrative. A number of subjects have demonstrated strong recourse to a recurrent theme, which represents something of a behavioural attractor to which they return again and again in their narrative, which suggests that this theme has some strong significance in their subjective experience of school mathematics. Examples would include Liam's repetitive reference to the notion of 'keeping up with the others'; Anna's apparently ever-present need to redefine herself from a 'D' to a 'B'; Scott's inescapable characterisation of his relationship to mathematics as 'struggle', and so on. Such dominant narratives seem to touch every aspect of affect, from emotion to attitudes, beliefs and even their identity as mathematical learners, thus demonstrating the complex and intimate relationships between affective constructs within their subjective experience.

Looked at in the light of motivational explanations, most of these examples appear to be rooted in the experience of the mastery state (usually mastery-losing),[3] and this lack of mastery appears to seep into every affective aspect of their relationship to learning the subject.

Another class of pattern in behaviour is that of engagement structures. A comprehensive account of emotions, and their relation to motivation and other affective variables in mathematics education is given by Goldin and colleagues (Goldin et al., 2011). One of the strengths of this approach is that it takes account of the variability of in-the-moment goals. There is an understanding that the engagement structures that we bring to situations in the mathematics classroom alter in real time according to the outcomes within the flow of our subjective experience. Thus the approach goes beyond trait considerations to ask not just 'are you disengaged', or even 'how much', but to ask the deeper questions of why and how you are disengaged. They argue that the immediate construed situation is more important in our moment-by-moment experience than trait. Goldin et al. have identified nine such structures that they propose may be universal in a mathematics classroom context. Examples include: 'Let me teach you', 'Look how smart I am', 'It's not fair', 'Don't disrespect me'. By focussing on data from the current study, evidence supports the existence of some of these structures, and suggests deeper, underlying motivational factors that bring them about.

These structures can be used as an interpretive lens which integrates different aspects of students experience as they mesh into semi-stable patterns of behaviour. The data here supports the notion of such engagement structures. The principles of Reversal Theory would lead us to predict that experiences or episodes would be reported and interpreted according to the motivational state experienced at that time, together with the associated values, feelings and emotions, which will play out in the behaviour of the student. Depending on perceived outcome, these will be experienced as positive or negative. To take an example, when mastery and other-oriented states are combined, a person will be able to identify with others, and will be focussed on vicarious power. In this way, supporting and helping others is seen as a universal motivational drive. The relation between motivational states and engagement structures can be illustrated by taking the example of the engagement structure '*Let me teach you.*' This structure is described as the need to help others by explaining or demonstrating. As noted above, in the card sort, the card 'helping others' was chosen by half of the students, and the notions 'Let me teach you' and 'helping others' appear to be equivalent. Students' accounts of their experiences, and their importance therefore serve as supportive evidence for the related structure *'Let me teach you.'*

There is also in this data evidence of other possible candidates for engagement structures. One such example is *'I want to have fun',* although there may be some overlap with the structure *'I'm really into this.'* Having fun, in motivational terms arises when the person is in the paratelic (playful) state. It is associated with immersion in a task or activity for its own sake, with spontaneity, taking risks and seeking arousal. It is associated with the emotion of excitement. It can be triggered by curiosity, intrigue, novelty, or by making an activity into a game. On the other hand, if someone is in the paratelic state and arousal is low, it will be experienced as boredom or sullenness. Discussion of this theme arises in the interview narrative from a number of sources. One source is when the emotion-card excitement/curiosity', or when the motivation-card 'having fun' is chosen (see for instance, Nora or Uma). It also arises in the narrative in response to a prompt to describe their best/favourite experiences of maths, or to projective questions like – 'what would you like maths lessons to be like?' (e.g., Mary or Mike below)

> If you're not enjoying the course, what's the point in doing it? Fun is.. doing something with friends…doing something new – that's what makes me feel good. (Nora)

> If I understand it I'll carry on with it…that's what makes it enjoyable…new techniques. (Mike)

> I like things which interest me. No excitement, no interest, don't learn. (Uma)

> Sometimes lessons are interesting…it tends to help my work. If I see it as a game I'll play it 'cos I like games. (Mary)

A natural question to ask is whether there are structures that militate against engagement – that characterise avoidance or outright disaffection. We might term these *disengagement structures*. The structures '*It's not fair*' and '*stay out of trouble*', identified by Goldin and colleagues, have some of these characteristics.

One such additional structure suggested by this data is *'You're not helping me/ you don't care.'* This structure is clearly related to Goldin's structure *'It's not fair'*, but it goes somewhat deeper in providing the motivational and emotional rationale for the genesis of that structure. In motivational terms it is self-sympathetic, with the associated negative emotion of resentment. The additional aspect of fairness (or lack of) also suggests negativism, where the negative emotion would be sullenness or anger.

In terms of evidence, it seems to be a common experience of disaffected students in this study to perceive that teachers pay less attention, and value less, students who are not fluent in mathematics, who are slower to develop understanding and competence, and, more specifically, students who will not achieve significant and valued exam results. The illustrative statements below tend to arise in discussing the 'dips' in life history ('me and mathematics'), but they also arise in other areas of the interview.

> I felt myself constantly being pushed down when I felt I should have been pushed up. A few of my teachers would focus a lot more on the ones who

> were getting it rather than the ones who needed help…and I found that really strange… I got dumped on somebody else…it's a horrible feeling. (Eve)

> (my decline)…was at the start of ability groups…you work it out for yourself (even though they don't say)…the teachers were focussed on the able pupils… it knocked my confidence…I can feel the teachers frustration at explaining over and over again…I need to do this myself because I'm not going to get any help. (Eve)

Finally, this evidence demonstrates that the notion of engagement structures posited by Goldin and colleagues has validity in drawing together evidence of affect from a variety of perspectives. They appear to offer an intermediate level construct that can encompass more fine-grained constructs such as motivation, emotions and beliefs, and at the same time provide a good fit with current data.

> …engagement structures help us to understand how achievement orientations, orientations and beliefs can influence the immediate, highly variable nature of students' behavioural, cognitive and affective engagement. (Goldin et al., 2011, p. 558)

MOTIVATIONAL PATHWAYS

A key feature of the evidence in this study is that motivational and emotional engagement in mathematics classroom results in frequent shifts of motivational state, and emotions result according to how the unfolding activity is perceived and interpreted by the person involved. The sequences described in this evidence are not random, and some explanation, based on a motivational and emotional interpretation, has been provided in a number of cases. Two contexts have emerged that seem to characterise much activity and learning in mathematics classrooms, and each of these involves a typical or predominant motivational pathway, which can be described here in more detail. The first of these can be described as the serious-conforming pathway. It is characterised in the table below:

A danger with this pathway is that if the task is seen as trivial or irrelevant, that is, the outcome becomes motivationally null, then a reversal to the paratelic state may occur, resulting in boredom.

Successful engagement and performance in this mode is maintained by metacognitive skills such as effort and persistence. The motivational 'hook' is about needing to do this activity for a reason. The optimal version of this mode can be summarised as 'mastery for a reason'. Developing a narrative of significance will enhance motivation in this mode, but at the same time, the route to successful mastery needs to be clear and perceived as possible. For students, explanation is seen as key, along with other pedagogical processes to enable relational understanding, leading to competence. Mathematics, along with any other area of human performance (such as sport) is seen as needing practice, drill, effort and persistence.

Table 14. Mastery for a reason versus 'what's the point?'

Nature of activity	*Mathematical context*	*Motivational state*	*Motivational requirement*	*Positive affective correlates – If successful*	*Negative affective correlates – If unsuccessful*
Procedural tasks	Procedure to be learnt and performed correctly. Speed and right answers rewarded	Telic Conforming Mastery	Telic – that meaning, purpose or utility are established. Conforming – sense of duty. 'Correct' procedure to be followed Mastery – I understand, so I can do	Need to apply effort to overcome negative feeling in order to learn. Performing correctly results in a sense of relaxation/relief. A sense of achievement (if it is perceived to be important, or if it took effort to learn). A sense of pride at overcoming and winning	Anxiety, fear or panic when faced with challenge. Possible switch to anger if effort is unsuccessful. Sense of detachment or being excluded in case of failure, resulting in mastery-losing and sense of humiliation

The mode of engagement is by far the most usual mode referenced in classes in this study, and reported by subjects. For some, it appears there seems to be little variation. Where it becomes institutionalised, it can become the dull 'textbook and worksheet diet' of drill and exercises described so vividly by many of the subjects of this study.

It is also worth pointing out that any possible pleasure and satisfaction here is deferred. In the telic state, this is only likely to be available on successful completion of a task. The satisfaction (e.g., relaxation/relief) is a low arousal emotion, and thus is unlikely to be experienced intensely. In a classroom context, it is also likely to be transitory, since as soon as the next task appears, the positive feeling is likely to fade into the background. The danger is that the satisfaction may be outweighed by the effort invested.

The second pathway or mode is the playful-rebellious route.

The need for fun appears so often in the literature and in this study, it seems clear that it provides important motivational satisfaction. It says something important about how these young people engage and connect with learning. The evidence here is that it is not trivial, and not an add-on that makes difficult tasks more palatable. Looked at from a motivational point of view, it is a motivationally totally different way of engaging people in learning mathematics. Good teachers have always known this. In other domains of life, such as sport, the need for paratelic engagement and

Table 15. Engagement and pleasure through interest versus 'I can't be bothered'

Nature of activity	*Mathematical context*	*Motivational state*	*Motivational requirements*	*Positive affective correlates – if successful*	*Negative affective correlates – if unsuccessful*
Open-ended task or problem to solve or investigate	Mathematical thinking and heuristic skills required. No necessary single right answer. Methods are undetermined and in control of student. Possibility of group working. Making connections, reasoning and understanding are privileged.	Paratelic Rebellious Mastery	Curiosity, real-world interest, challenge, game, unfamiliarity or novelty. Activation of exploration and discovery. Need to share within the group	Fun, excitement, enjoyment and immersion in the task at hand. Sympathetic satisfaction from working in a group.	Boredom if the challenge is too easy. If too hard, a reversal to telic could induce anxiety, or shift of focus to mastery-losing resulting in humiliation

enjoyment, even in training and practise situations, is well understood and used in a structured way to be incorporated into programmes and routines.

Yet the mathematical education community has rarely had the language or the theoretical frameworks to discuss its importance. It has thus been trivialised as 'fun' and so relegated in importance. It is seen as antithetical to many of the established pedagogical practices of many classes. In many classrooms described in this study, it is shut off as a motivational opportunity. Yet at the same time, we have also presented evidence that students will often be in the playful state. There needs to be pedagogically sound and justified ways of incorporating this into the life of the classroom.

It is often assumed in the literature (as pointed out in the review of literature) that educational activity is goal-driven. Reversal Theory contradicts this and suggests that although activity and directed experience seeks to satisfy needs, these needs are not always expressed as goals. When people are in the paratelic state, goals and outcomes are not in the phenomenal frame. People engage in activity for the momentary pleasure it affords them. This is actually a desirable state in which to engage with mathematical contexts. The desire to engage and involve oneself in a mathematical activity without paying attention to or being driven by destination releases a person and allows them to follow their interest and instinct; to look for patterns and insights. It enables creativity and pattern-spotting. It is precisely where the 'rich learning' described in ACME (2012) comes from. Indeed it is possible to go further and to say that to be paratelically engaged with mathematics is a sign of affection for mathematics, since our interest and curiosity will enable us to engage without fear. The inability to do so is a litmus test of disaffection.

Further, the poverty of mathematics curricula and pedagogy is precisely that paratelic activity and engagement is not encouraged and enabled in so many mathematics classrooms.

MOTIVATIONAL CLIMATE FOR LEARNING MATHEMATICS

In looking at the influence of aspects of affect on learning and achievement in mathematics, a range of constructs have been examined and evidenced. Many of these have provided evidence of the bi-directional relationships between affect, cognition and learning. However, here, I would like to examine the notion of classroom climate, which has been identified as having potential to offer further insights into the relationship between affect and achievement.

By analysing the complex and dynamic interactions between aspects of affect (primarily motivation and emotion) and how these influence how young people do or do not learn mathematics, I will propose a framework for understanding the notion of classroom climate as an intersubjective space with psychosocial, individual and educational aspects, where the influence of classroom practices and conditions are brought to bear on learning. By understanding how the experience of learning mathematics for the individual can be influenced by the way that teachers create a positive climate for learning, it may be possible to understand how the decline into disaffection might be alleviated.

Classroom Climate

Evidence shows that classroom climate is important when accounting for wide variations in measures of affect and attainment between groups in the same institution. Noyes (2012) has noted significant within-school variation in teacher quality, student engagement and learning outcomes. He cites Opdennakker et al. (2002), who concludes that "the most important differences between teachers (with respect to explaining differences in mathematics achievement) exist in their effect on the learning climate in the class" (p. 418).

Noyes notes in relation to this variation that students seem to have a very different pedagogical experience depending on which group they are in and which teacher they have. He goes on to note the importance of student-centred curriculum and pedagogy in the development of positive attitude and disposition towards mathematics.

In chapter 5, evidence is presented of a survey of a whole school year 9 (aged 13–14 years) cohort in the UK. Since the groups were organised by attainment level in two parallel half years, it was possible to look at variations of aspects of affect by group, including its relation to levels of attainment. In that study, measures of affect appear not to decline according to level of attainment. But since parallel groups at the same attainment level varied strongly on measures of affect, this suggests that it is the classroom or group itself that is the major determinant of pupils' affective experience of mathematics. Positive and negative affect, and perceived competence in mathematics do not appear to be related to the level at which pupils are achieving. Students in higher groups appear to be as likely to report negative affect as those in lower groups. The study also showed how widespread the experience of negative affect is, with boredom being particularly prevalent in the experience of these young people.

This evidence points to the idea that classroom climate seems to be an important potential area for study in terms of differences in students' experience of school mathematics, and its influence on learning.

Ambrose et al. (2010) define classroom climate as:

> the intellectual, social, emotional, and physical environments in which our students learn. (p. 170)

There is less evidence of research on classroom climate in relation to learning of school mathematics. One such example is that of Goh et al. (1995), who talk about the influence of psychosocial climate on student outcomes in mathematics classrooms, and offer the four dimensions of climate as cohesion, competition, friction and task focus. They report:

> Research using student perceptions of classroom environment tends to support the contention that classroom learning environment could be an important variable accounting for variance in student outcomes. (p. 29)

That students need to feel supported has also been extensively examined in the classroom environment literature. Patrick, Ryan, and Kaplan (2007) suggest that the association between perceived classroom environment and student engagement is mediated by students' motivational beliefs.

Skovsmose (2005) talks about barriers to learning, and he states:

> a particular situation or a particular way of organising teaching-learning processes can prevent students from acting as learners. (p. 7)

He goes on to describe climate as:

> the opportunities which the social, political and cultural situation provides for the person. (p. 6)

In contrast to others, he notes the influence, not just of the past (e.g., inherited culture) but also of intention, or the individuals interpretation of the current and future opportunities, which he terms foreground. So, whilst acknowledging the focus on socio-cultural issues, it also seems that an individual's response to climate involves aspects of affect, including motivation and emotion.

Hannula (2012) reports that the 'microculture' of the classroom may also build resilience against overall educational policy. For example, classroom culture, interpreted by him within the context of motivational theory as involving community, autonomy and mastery goal orientation, has been found to mitigate the influence of the prevalent performance pressure in the U.S. educational system. He goes on to report the work of Turner, Meyer and Schweinle (2003) that studies on classroom affect and motivation often emphasise the teacher role in the establishment of classroom discourse. Patrick et al. (2007) also point out that the association between perceived classroom environment and student engagement is mediated by students' motivational beliefs.

Two conclusions can be drawn from this evidence. The first is that classroom climate can be seen to be an important factor in determining students' experience of school mathematics, in terms of affect, behaviour and achievement. Secondly, although issues of policy, physical environment, resources and the socio-cultural and socio-economic make-up of both schools and classrooms are deemed to be important, it also seems clear that the psycho-social and socio-emotional (i.e., the affective) climate within a classroom is an important consideration influencing students experience of, and engagement with learning.

The Role of the Teacher

If climate can vary significantly, when school climate can be said to apply and influence equally across those classrooms, then the difference is most likely to be with the individual teacher. To put it another way, the individual teacher can be seen to be the primary architect of classroom climate. This raises questions about what those teachers who create positive learning climates in their classrooms do, that is different to other teachers. Clearly, part of the answer is likely to be around pedagogical and teaching practises, but evidence also suggests that the personality of the teacher, and the nature of the interpersonal dynamics of the classroom are also relevant. Gossman (2011) in a study of reports entitled 'My best teacher', identified two separable categories labelled the 'teacher-as-teacher' and 'teacher-as-person'. Ryan and Patrick (2001) showed that where pupils viewed their teachers as supportive and willing to help them, they were more engaged in learning and less likely to be off task or disruptive. Troman et al. (2007) refer to the psychic rewards for both teachers and pupils, who may gain immense satisfaction from teacher-pupil relationships. These findings are echoed in Lewis and Forsythe (2012), which showed that a range of personal attributes such as enthusiasm and passion, and human/caring qualities were highly valued by students. That study also found, like Gossman (2011), that certain aspects of the teacher-as-teacher role helped to create a positive and engaging learning environment. These included real world interest, clarity of explanations, fun, and the use of a variety of methods and representations, combined with the ability to provide a sense of order and discipline, and the right level of challenge. Students also reported being very influenced by aspects of the teacher's personality and approach, including energy, passion and enthusiasm (in general, but also specifically for mathematics); the human and caring aspect of the relationship which built confidence in students.

Some deeper principles appear to be in operation in the classrooms reported by Watson and de Geest (2005). They suggest that a range of beliefs and commitments that underpinned teacher choices are more important than individual pedagogical practices in fostering learning. That study showed that low attaining pupils can develop skills to think mathematically. It also showed that in doing so, pupils outperformed comparison groups, particularly on test items that involved mathematical thinking. This was achieved in part by emphasising choice, freedom, challenge, and devolved responsibility on mathematically challenging tasks over extended periods of time. The valuing of students both as learners and as individuals seemed to be important, and Watson and de Geest (2005) talk of 'togetherness' and 'humane attitudes', emphasising the social and human aspects of the classroom environment.

This evidence in relation to classroom climate, and the role of the individual teacher, will be reflected in the way that climate will be characterised here.

Qualitative evidence has been put forward to show how affective factors manifest in the felt experience of young people, and how they interpret and respond to the availability or lack of motivational affordances in the climate of the classroom, and how these in turn enable or limit the opportunities to learn mathematics.

It has been noted, for instance, that a change of teacher can be the primary reason for both dips in the life-histories of students, but also in the recovery phases. This indicates that the individual teacher is the prime architect of the experience of learning (or not learning) mathematics in the classroom apart from the student themself. If that is so, it makes sense to look at the evidence of the influence of classroom climate, teacher's personality and teaching style on the students' experience of mathematics.

Whilst socio-cultural context is so often defined in terms of power relations, they can also be conceived of in motivational terms. Indeed, Skovsmose defines the notion of foreground in terms of the meanings that an individual interprets about his/her context. Teaching style can be thought of as the collection of pedagogical habits, routines and approaches that typify the way a teacher teaches. That the personality of the teacher is important in creating climate, influencing teaching styles, is evidenced consistently here. Further, the personality of the individual teacher is shown to influence the relationship with the student, and therefore the climate for learning.

There is a great deal of evidence here of the positive aspects of teachers as personalities, and of positive classroom climates. In addition to the evidence embedded in the discussion above, one further source has some interesting perspectives on this issue. Lewis and Forsythe (2012) examined 75 written texts (from two separate cohorts) of prospective mathematics PGCE students, who wrote about a mathematics teacher that they admired. The literature review for this paper identified classroom climate as an important factor influencing students learning, and in particular that the emotional dimension of the relationship between teacher and student was important. Surprisingly, given that these students were successful in their school mathematics careers, and were aspiring to be teachers of mathematics, a number reported periods of disaffection. Being ignored or feeling unsupported were key aspects of this disaffection.

Students reported being very influenced by aspects of the teacher's personality and approach. These included energy, passion and enthusiasm (in general, but also for mathematics); the human and caring aspect of the relationship which built confidence in students. In terms of pedagogy and teaching approach, a number of students mentioned an interest in relating mathematics to real world issues; the importance of clarity and simplicity in explanations; the judicious use of a variety of methods, artefacts and representations; and inevitably, fun and a sense of humour.

A further aspect of admiration was the setting of a disciplined climate where students could relax and work without friction or distraction. For instance, distraction by other pupils, particularly if they are misbehaving, is cited as a negative (or disaffecting) feature of many classrooms. Pupils and students seem to have a low tolerance for teachers who can't control classrooms, and value order and structure.

> it's a really big distraction which affected my education as well cos…she didn't really give attention to ones who actually do wanna learn…and paid more attention to those who were messing around…and trying to get them to learn (Adnan)

Teachers who don't explain, and expect pupils to work it out for themselves are criticised, as are teachers who spend more time with disruptive pupils, or with those who are seen to be 'clever'.

> because the teachers…they didn't explain properly innit…they didn't say how to work it out…they just give you worksheets…and you do it and they mark it…(Nina)

There is much evidence here that pupils don't learn from books alone, and don't enjoy classrooms where books are the main medium of activity.

> mmm…yeh…I did enjoy it…cos we used to have a lot of…for example we weren't just sitting there and writing in books…we used to have…like counters…and different things…like visual…things you could touch…which would help you…more like real life situations…that you'd actually use…so it was interesting…you weren't just sitting there writing in a book…and learning that way…(Adnan)

> yeh…I can't…just sit there and write in a book…and there were times in year 4 to year 5 and we had to read…out of books and write…read and write…and that's all it was…I did not like that…I prefer more practical work…learning in a practical way…you understand more…you don't just look there and do the work…you actually get to experience (Adnan)

Mathematics classrooms, the evidence here suggests, are focussed, not surprisingly, on competence – the development of competence and the demonstration of that competence. However, the evidence here offers some balance to the exclusive focus on competence (mastery), to paint a picture in which the sympathetic motivational state plays a key part. Students clearly bring their emotional and human selves to bear in mathematics classrooms, yet it is little taken into account in much research, which has 'cold' cognition as its focus. Thus we see evidence that students can be disaffected because they feel uncared for by teachers. We see that where their full emotional and human selves are ignored, the classroom experience will be impoverished.

It has already been noted, for instance, that many teachers appear not to know or understand the rich motivational value that there is in creating an environment where pupils/students can help others and be helped in turn. The motivational case for cooperative collaborative working is evidenced strongly in this study.

The evidence in this study has shown that students can be motivated (by which is meant to be motivationally connected) in a whole range of ways to mathematics. The evidence has shown that although disaffected students seem to find motivational connection with some aspects of the work in their classrooms, the reason they are disaffected is that some of these routes to satisfaction appear to be blocked off or otherwise unavailable. So, although we see much evidence here of motivationally impoverished mathematics classrooms, it does lead to the question of what a motivationally rich mathematics classroom might look like.

Motivationally Rich Climate for Learning Mathematics

At a very general level, the answer appears to be simple: it is a classroom environment where students have the opportunity to gain motivational satisfaction in any motivational state by undertaking mathematically rich activities. If there are motivational opportunities for satisfaction in every state, then this means that students can be connected to the activities of the classroom, whatever particular state they are in at any moment. In practical terms, this may be impossible. However, it is not unreasonable to make two suggestions. Firstly, although opportunities may not be available in all states at all times, in a motivationally rich environment, opportunities to gain motivational connection in every state will be available over time. Secondly, this leads to the suggestion that a motivationally rich mathematics classroom is one in which the opportunities to engage with the mathematics in qualitatively different and satisfying ways is maximised.

A good comparison is with the notion of 'diet'. In a motivationally rich mathematics classroom all opportunities are available – although not necessary all at the same time.

One way a model of such a climate can be constructed is to look at the needs suggested by each motivational state (or, for transactional states, combinations of states), and to match this to legitimate and motivationally rich activities and characteristics in a mathematics classroom. How it might work is illustrated below.

Few of the suggestions in the right hand column are new or revolutionary. Indeed, part of their power is that they are well established and evidenced in the literature.

Table 16. Motivational climate for learning mathematics

State	*Needs*	*Opportunity*
serious	Significance achievement	The importance and value of mathematics is explored and explained Mathematics is seen to be relevant (utility) in the real world Reasons and rationales for topics and activities are explained Immediate and longer term targets are seen as meaningful
playful	Fun excitement	Opportunities are available to explore Activities are framed as games Intrigue and curiosity are used A variety of methods and representations is used
conforming	Inclusion, feeling part of the group	Expectations are set for mathematical outcomes and behaviour, habits All students are respected equally
rebellious	Individuality, freedom	There is scope for individuality in approach Opportunities to develop ‘own methods’ are available
self-mastery	Competence, power, control	Explanations and opportunities are provided for students to understand A focus on relational and not just procedural knowledge Success is built into activities Students must feel they CAN DO what is required Persistence and effort are framed as normal and necessary
self-sympathy	Feeling cared for	The teacher is seen as human Students feel cared for
other mastery	To help others	Group working is encouraged and enabled Helping each other is the norm
other sympathy	To care for others	The mathematics class is seen as a social environment

What is of interest here is two things. Firstly, they are grown outwards from the notion of motivation. They thus have explanatory power. By conceiving of them in this way, we can understand why they are important. Secondly, they have a power by virtue of being held in a single, theoretic framework.

NOTES

[1] It seems logical that people could move in either direction along this space.

[2] I discussed this issue at length with Peter Liljedahl at Cerme 8, who has extensive experience of change in teachers and classrooms, and he is confident there is much evidence of genuine sustainable re-affection in students. However, at present, this evidence is anecdotal and not yet written up.

[3] The exception being a dominant narrative around the notion 'nobody cares for me'.

CHAPTER 15

PERSPECTIVES ON RESEARCH INTO DISAFFECTION

Inevitably, with any programme of research, at the conclusion of the research, there remain as many questions unanswered as there are answered. In this final chapter, I will review the outcomes of the research, and try to place them in context, and, in particular, to look forward to the directions in which future research might take us.

CHALLENGING ASSUMPTIONS

The research has brought forward evidence that runs counter to some of the assumptions that are both explicitly and implicitly often taken to be true by researchers in the field. In this section, I will present a number of these.

Stability of Dispositions Towards Mathematics

It was pointed out in earlier chapters that there has been a tendency in research to assume that dispositions towards school mathematics, once crystalised as negative, tend to remain, or even harden. The evidence here, and particularly that of the life histories, has shown that young peoples' relationship to school mathematics is fluid and quite volatile across the years. Once aware of this ebb and flow, it becomes less prudent to treat such dispositions as trait, and this in turn should discourage us from methods and approaches to researching affect that are based on such an assumption.

There are a range of factors that account for the 'ups and down' in these relationships, but the most clearly and consistently cited factor in the changes in relationship is the influence of the teacher. In this study, this influence has been presented as sensitivity to classroom climate, and specifically as characterised by motivational and emotional factors. Certainly, by re-thinking of affect (or, more specifically, of disaffection) in this way, opens up the possibility that it is possible to recover from a disaffected disposition. It has been proposed here that disaffection leads to disablement of learning, but disablement does not mean incapable. I propose that the possibility of re-affection, and the conditions of such re-affection be taken seriously as an important phenomenon to study.

Young Disaffected People Are not Articulate about Their Own Affective Landscape

It was also noted in earlier chapters the difficulties of obtaining high quality data relating to affective issues. Part of this difficulty is seen to be that issues of attitude, motivation, beliefs and emotion are inherently difficult to articulate. However, there are also suggestions that disaffected or disenfranchised young people may not be easily be able to describe their own affective landscape with any depth, sophistication or clarity. I have found this not to be the case. In the interviews that I conducted (48 in all), I would judge that only one was inarticulate to the degree that it gave little insight into the subject's affective landscape.[1] On the contrary, I would suggest that the case studies especially illustrate that even highly disaffected students can be highly self-aware and articulate. The exception to this, as has been pointed out, is the small but significant minority of respondents reported in chapter 5 who seemed unable to recognise the four negative transactional emotions.

It should be acknowledged that the use of innovative methods and instruments not only provide good data in their own right, but also provide the stimulus, prompt and the occasion within the context of the interview, to discuss complex issues. It seems that structured, visual and projective techniques can act as midwife to deeper thoughts and as a window into our deeper selves.

Self Regulatory Skills

In a similar vein, it is often assumed or stated explicitly that students who achieve only at a low level have less well developed metacognitive or self-regulatory skills. As has been pointed out, the evidence here suggests that that many of these highly disaffected young people apply high levels of effort and determination, even when it is painful for them to do so. Evidence has also been presented that students are often very thoughtful in the way they manage their own negative emotions – particularly anger.

Equally, unsuccessful or disaffected learners, such as many in this study, will be predicted to have a 'performance' (as opposed to a 'mastery') mindset. No simple binary distinction was observed in this study. Some of the most disaffected students (such as Anna) were quite clearly aware of the value of effort and hard work in learning, even as they expressed strong needs to perform well in relation to others. Put another way, they are not fixed in a single mindset (as in trait), but seem to operate at times in both mastery and performance modes.

USE OF REVERSAL THEORY

A case has been made that theoretical innovation was necessary to investigate more thoroughly the existence, genesis, and phenomenology of affect in mathematics education beyond quantitative descriptions of attitude. On that basis, Reversal Theory was adopted as an alternative theoretical framework for the study of affect in mathematics education.

It was not viewed as an aim of this study to either confirm or disconfirm the theory, or to find evidence to test hypotheses based on the theory, over and above the reasonable expectation that the existence of motivational states and their associated emotions would somehow play out in the experience of young people's relationships to mathematics. The theory was seen as useful in three main areas. The first of these was as a framework that informed the design and development of instruments and methods for capturing data. In particular, the theory offered a comprehensive and coherent account of 16 primary emotions. Theoretical frameworks accounting for emotions are few and far between in mathematics education research, and theoretical validity is required in the field. Reversal Theory not only provides such a structure, it enables us to give an explanatory account of how and why emotions arise in educational settings. The Reversal Theory account of emotions is not contradictory to that of Pekrun et al. but adds a theoretical underpinning and explanatory value. The two dimensional arrangement of emotions in Pekrun et al. is given a more full explanation in the Reversal Theory account. In a similar way that the theory informed the use of the TESI-ME to survey experiences of negative emotions, the theory was also used to inform the design of the cards, which elicited rich data about positive experiences of school mathematics.

The second use of the theory was as a framework that could inform the interpretation of data. The theory was used for more than just categorising data segments. Using the theory, it has been possible to interpret episodes as sequences of experience through the lens of the motivational states, and associated concepts such as reversal and shift of focus. So, not only has the theory been useful in this way in interpreting students accounts of their experiences in mathematics classrooms, it has provided further evidence both of the states themselves and how they play out in the educational experiences of the young people in this study, but also as a further confirmation of concepts such as reversal.

It can reasonably be claimed that by looking at episodes through the motivational and emotional lens afforded by the theory, it has been possible to provide an explanatory commentary as to what is happening in the mathematical experience of these young people. It has been shown that the theory can also provide an underpinning explanatory framework for other constructs in the field, that are evidenced in research, but have, as yet, little theoretical basis. The notion of engagement structures (Goldin et al., 2011) comes into this category, since an account has been offered showing the motivational and emotional basis for a number of these. It is possible that the theory could in future be used further in this way to provide a theoretical and explanatory basis for other phenomena and models.

Alongside this, it has been demonstrated that the theory can provide the basis for a critique of other theories in the field. Such a critique has been offered here of Self Determination theory, and might provide explanations, for instance, as to why there are variable results of predictions made on the basis of that theory.

A third way in which the theory has been used is as a basis for theorising around factors that influence the motivation of young people to learn mathematics. The notion of motivational climate for learning mathematics has been proposed and introduced. This theorising is consistent with the data in the study, but it provides a radically new kind of framework with which to understand what happens in a mathematics classroom. Its status, then, is as a proposed framework, but one that has potential to be confirmed in further study, and one which can provide new insights and explanations in the future.

Taken as a whole, these three complementary aspects of the use of the theory can be judged as a justification for the adoption of the theory within the context of this study. It also suggests that the theory has potential to inform further research in the area.

A Final Note

This research project was undertaken to investigate and to understand better the landscape of disaffection with school mathematics. I believe that it has met that aim, and that a number of themes have been drawn out as worthy of further research. I would suggest that first amongst these is the notion of motivational climate for learning mathematics, and it's potential role in the re-affection of disaffected young people.

One further conclusion that can reasonably be drawn relates to the implications for mathematics teacher education. Since much research activity has been placed on the notion that teachers should have mathematical and pedagogical knowledge, I would claim that they should also have what might be termed motivational or affective literacy. By this is meant that an awareness of the importance of motivational, emotional and other affective factors is vital for any successful teacher of mathematics. At the moment, attention to such motivational aspects and their impact on learning (or not learning) mathematics seems to play little part in the education of teachers. The results presented here have many lessons for teachers, policy makers and other stakeholders involved in the design and delivery of mathematics curricula, and materials to support the learning of mathematics.

NOTE

[1] That interview was with Sam, whose narrative of his relationship to mathematics was almost exclusively centred on the single word 'struggle'.

REFERENCES

Adamson, P. (2007). *Child poverty in perspective: An overview of child well-being in rich countries.* Report Card 7. Florence, Italy: UNICEF Innocenti Research Centre.

Allen, B., D., & Carifio, J. (1995). The relationship between emotional states and solving complex problems. *Eastern Educational Research Association.*

Ambrose, S. A., Bridges, M. W., DiPietro, M., & Lovett, M. C. (2010). *How learning works: Seven research-based principles for smart teaching.* San Francisco, CA: Jossey Bass.

Anderman, E. M., & Wolters, C. A. (2006). Goals, values and affect: Influences on student motivation. In P. Alexander & P. Winne (Ed.), *Handbook of educational psychology* (pp. 369–389). Mahwah, NJ: Erlbaum.

Apter, M. J. (Ed.). (2001). *Motivational styles in everyday life: A guide to reversal theory.* Washington, DC: American Psychological Association.

Apter, M. J. (2013). Developing reversal theory: Some suggestions for future research. *Journal of Motivation, Emotion, and Personality, 1,* 1–8.

Birdwell, J., Grist, M., & Margo, J. (2011). *The fogotten half.* London, England: Demos.

Boaler, J. (2009). *Experiencing school mathematics.* Buckingham, England: Open University Press.

Boaler, J., William, D., & Brown, M. (2000). 'Students' experiences of ability grouping – Disaffection, polarisation and the construction of failure. *British Educational Research Journal, 26*(5), 631–648.

Borthwick, A. (2011). Children's perceptions of, and attitudes towards, their mathematics lessons. In C. Smith (Ed.), *Proceedings of the British Society for Research into Learning Mathematics, 31*(1).

Breen, C. (2000) Becoming more aware: Psychoanalytic insights concerning fear and relationship in the mathematics classroom. *Proceedings of the 24th Conference of the International Group for the Psychology of Mathematics Education (PME)* (Vol. 2, pp. 105–112). Hiroshima, Japan.

Brown, L., & Reid, D. A., (2006). Embodied cognition: Somatic markers, purposes and emotional orientations. *Educational Studies in Mathematics, 63,* 179–192.

Brown, M., Brown, P., & Bibby, T. (2007). I would rather die: Attitudes of 16 year-olds towards their future participation in mathematics. In D. Küchemann (Ed.), *Proceedings of the British Society for Research into Learning Mathematics, 27*(10), 18–23.

Buxton, L. (1981). *Do you panic about maths: Coping with maths anxiety.* London, England: Heinemann Educational.

Caine, G., & Caine, R. N. (2006). Meaningful learning and the executive functions of the brain. *New Directions for Adult and Continuing Education, 110,* 53–72.

Cameron, J., & Pierce, W. D., (1994). Reinforcement, reward, and intrinsic motivation: A meta-analysis. *Review of Educational Research, 64,* 363–423.

CBI. (2010). *Making it all add up: Business priorities for numeracy and maths.* London, England: CBI. Retrieved on November, 2012 from http://www.cbi.org.uk/media/935352/2010.08-making-it-all-add-up.pdf

Chula, M. (1998). *Adolescents' drawings: A view of their worlds.* Paper presented at the Annual Meeting of the American Educational Research Association. San Diego, CA

Cockcroft, W. (1982). *The Cockcroft report: Mathematics counts.* London, England: HMSO.

Coffield, F., Mosely, D., & Ecclestone, K. (2004). *Learning styles: What theory has to say about practice.* London, England: Learning and Skills Research Centre.

Cohen, L., Manion, L., & Morrison, K. (2000). *Research methods in education* (5th ed.). London, England: RoutledgeFalmer.

Cramer, K. (2011). Six criteria for a viable theory: Putting reversal theory to the test. In *Proceedings of the 15th International Reversal Theory Conference,* Alexandria, VA.

Croghan, R., Griffin, C., Hunter, J., & Pheonix, A. (2008). Young people's construction of self: Notes of the use and analysis of the photo-elicitation methods. *International Journal of Social Research Methodology, 11*(4), 345–356.

REFERENCES

Csikszentmihalyi, M. (1996). *Creativity: Flow and the psychology of discovery and invention.* New York, NY: HarperPerennial.

Davidson, J., Dottin, J., Penna, S., & Robertson, S. (2009). Visual sources and the qualitative research dissertation: Ethics, evidence and the politics of academia – Moving innovation in higher education from the center to the margin. *International Journal of Education and the Arts, 10*(27).

DeBellis, V. A., & Goldin, G. A., (2006). Affect and meta-affect in mathematical problem solving: A representational perspective. *Educational Studies in Mathematics, 63*(2), 131–147.

Deci, E. L. (1971). Effects of externally mediated rewards on intrinsic motivation. *Journal of Personality and Social Psychology, 18,* 105–115.

Deci, E. L., Koestner, R., & Ryan, R. M. (2001a). Extrinsic rewards and intrinsic motivation in education: Reconsidered once again. *Review of Educational Research, 71*(1), 1–27.

Deci, E. L., Ryan, R. M., & Koestner, R. (2001b). The pervasive negative effects of rewards on intrinsic motivation: Response to Cameron. *Review of Educational Research, 71*(1), 43–51.

Donaldson, M. (1978). *Children's minds.* Glasgow, Scotland: Fontana Original.

Dweck, C. S. (2000). *Self-theories: Their role in motivation, personality and development.* Hove, England: Psychology Press.

Early, R. E. (1992). The alchemy of mathematical experience:A psychoanalysis of student writings. *For the Learning of Mathematics, 12*(1), 15–20.

Evans, J., Morgan, C., & Tsatsaroni, A. (2006). Discursive positioning and emotion in school mathematics practices. *Educational Studies in Mathematics, 63,* 209–226.

Gee, P. (2011). Approach and sensibility: A personal reflection on analysis and writing using interpretative phenomenological analysis. *Qualitative Methods in Psychology Bulletin B, 11*, 8–22.

Givvin, K. B., Stipek, D. J., Salmon, J. M., & MacGyvers, V. L. (1996). *Teachers understanding of their students' motivation.* American Educational Research Association.

Goh, S. C., Young, D. J., & Fraser, B. J. (1995). Psychosocial climate and student outcomes in elementary mathematics classrooms: A multilevel analysis. *Journal of Experimental Education, 64*(1), 29–40.

Goldin, G. (2003). Developing complex understandings: On the relation of mathematics education research to mathematics. *Educational Studies in Mathematics, 54,* 171–202.

Goldin, G. A., Epstein, Y. M., Schorr, R. Y., & Warner, L. B. (2011). Beliefs and engagement structures: Behind the affective dimension of mathematical learning. *ZDM Mathematics Education, 43,* 547–560.

Gorard, S., See, B., & Davies, P. (2011). Attitudes, aspirations and behaviour in educational attainment: Exploring causality. *Bera Conference.*

Gossman, P. (2011). My best teacher. *Tean Journal, 3*(1).

Gross, J. (2009). *Long term costs of numeracy difficulties.* London, England: Every Child a Chance Trust.

Hannula, M. S. (2002). Attitude towards mathematics: Emotions, expectations and values. *Educational Studies in Mathematics, 49*(1), 25–46.

Hannula, M. S. (2004). *Affect in mathematical thinking and learning* (Ph.D. Thesis). Finland: University of Turku.

Hannula, M. S. (2006). Motivation in mathematics: Goals reflected in emotions. *Educational Studies in Mathematics, 63*(2), 165–178.

Hannula, M. S. (2012). Exploring new dimensions of mathematics-related affect: Embodied and social theories. *Research in Mathematics Education, 14*(2).

Hannula, M. S., Pantziara, M., Waege, K., & Schloglmann, W. (2009). *Multimethod approaches to the multidimensional affect in mathematics education.* Paper presented at Proceedings of CERME 6, Working Group 1.

Hembree, R. (1990). The nature, effects, and relief of mathematics anxiety. *Journal for Research in Mathematics Education, 21*(1), 33–46.

Hidi, S., & Harackiewicz, J. (2000). Motivating the academically unmotivated: A critical issue for the 21st century. *Review of Educational Research, 70*(2), 151–179.

Hodkinson, P., & Macleod, F. (2010). Contrasting concepts of learning and contrasting research methodologies: Affinities and bias. *British Educational Research Journal, 36*(2), 173–189.

Kerr, J. H. (Ed.). (1999). *Experiencing sport: Reversal theory.* Chichester, England: Wiley.

Kinder, K., Wakefield, A., & Wilkin, A. (1996). *Talking back: Pupil views on disaffection.* Slough, England: NFER.

Lafreniere, K. D., Ledgerwood, D. M., & Murgatroyd, S. J. (2001). Psychopathology, therapy, and counselling. In M. J. Apter (Ed.), *Motivational styles in everyday life: A guide to reversal theory* (pp. 263–286). Washington, DC: American Psychological Association.

Leder, G., & Forgasz, H. (2006). Affect and mathematics education. In A. Gutierrez & P. Boero (Eds.), *Handbook of research on the psychology of mathematics education, past, present, future.* England: Sense Publishers.

Lewis, G. (2011). The poverty of motivation: A study of disaffection with school mathematics. In *Proceedings of the 35th Conference of the International Group for the Psychology of Mathematics Education* (pp. 137–144).

Lewis, G., & Forsythe, S. (2012). Which qualities did aspiring teachers value in their 'best' mathematics teachers? *Teachers Education Advancement Network Journal, 4*(3), 46–58.

Lumby, J. (2012). Disengaged and disaffected young people: Surviving the system. *British Education Research Journal, 38*(2), 261–279.

Ma, X., & Kishor, N. (1997). Assessing the relationship between attitude towards mathematics and achievement in mathematics, a meta-analysis. *Journal for Research in Mathematics Education, 28*(1), 26–47.

Mackenzie, N., & Knipe, S. (2006). Research dilemmas: Paradigms, methods and methodology. *Issues in Educational Research,16*(2), 193–205. (Accessed September 30, 2013)

Mallows, D. (2007). *Switch to better behaviour management.* Dereham, England: Peter Francis.

Malmivuori, M. L. (2006). Affect and self-regulation. *Educational Studies in Mathematics, 63*(2), 149–164.

Mathews, A., & Pepper, D. (2005). *Evaluation of participation in A-level mathematics.* London, England: QCA.

McLeod, D. (1992). Research on affect in mathematics education: A reconceptualization. In D. A. Grouws (Ed.), *Handbook of research on mathematics teaching and learning* (pp. 575–598). New York, NY: NCTM.

McLeod, D. (1994). Research on affect and mathematics learning in the JRME: 1970 to the present. *Journal for Research in Mathematics Education, 25*(6), 637–647.

Meyer, D. K., & Turner, J. C., (2010). Discovering emotion in classroom motivation research. *Educational Psychologist, 37*(2), 107–114.

Middleton, J. A., & Spanias, P. A. (1999). Motivation for achievement in mathematics: Findings, generalizations, and criticisms of the research. *Journal for Research in Mathematics Education, 30*(1), 65–88.

Nardi, E., & Steward, S. (2003). Is mathematics T.I.R.E.D.? A profile of quiet disaffection in the secondary mathematics classroom. *British Educational Research Journal, 29*(3), 345–367.

Niemiec, C. P., & Ryan, R. M. (2009). Autonomy, competence, and relatedness in the classroom: Applying self-determination theory to educational practice. *Theory and Research in Education, 7*(2), 133–144.

Noss, A. (2009). *Disengagement from education among 14–16 year olds* (RR 178). National Centre for Social Research.

Noyes, A. (2012). It matters which class you are in: Student-centred teaching and the enjoyment of learning mathematics. *Research in Mathematics Education, 14*(3), 273–290.

Nussbaum, M. C. (1997). Capabilities and human rights. *Fordham Law Review, 66*(2), 273.

O'Connell, K. A., Gerkovich, M. M., & Cook, M. R. (1997). Relapse crises during smoking cessation. In S. Svebak & M. J. Apter (Eds.), *Stress and health: A reversal theory perspective.* Washington, DC: Taylor & Francis.

Opdenakker, M. C., van Damme, J., de Fraine, B., Landeghem, G., & Onghena, P. (2002). The effect of schools and classes on mathematics achievement. *School Effectiveness and School Improvement, 13*(4), 399–427.

Op't Eynde, P., de Corte, E., & Verschaffel, L. (2006). Accepting emotional complexity: A socio-constructivist perspective on the role of emotions in the mathematics classroom. *Educational Studies in Mathematics, 63,* 193–207.

Patrick, H., Ryan, A., & Kaplan, A. (2007). Early adolescents' perceptions of the classroom social environment, motivational beliefs, and engagement. *Journal of Educational Psychology, 99,* 83–98.

Pekrun, R., Frenzel, A. C., Goetz, T., & Perry, R. P. (2007). The control-value theory of achievement emotions: An integrative approach to emotions in education. In P. Schutz & R. Pekrun (Eds.), *Emotions in education* (pp. 13–36). Burlington, MA: Academic Press.

Philipp, R. A. (2007). Mathematics teachers' beliefs and affect. In F. K. Lester (Ed.), *Second handbook of research on mathematics teaching and learning* (pp. 257–318). Charlotte, NC: NCTM.

Pink, D. H. (2010). *Drive: The surprising truth about what motivates us.* Edinburgh, Scotland: Canongate Books.

Pintrich, P. (2003). A motivational science perspective on the role of student motivation in learning and teaching contexts. *Journal of Educational Psychology, 95*(4), 667–686.

Purkey, W. W., & Schmidt, J. (1987). *The inviting relationship: An expanded perspective for counselling.* Englewood Hills, NJ: Prentice Hall.

Rudduck, J. (2002). The transformative potential of consulting young people about teaching, learning and schooling. *Scottish Educational Review, 34*(2), 123–137.

Ruffell, M., Mason, J., & Allen, B. (1998) Studying attitude to mathematics. *Educational Studies in Mathematics, 35,* 1–18.

Schorr, R. Y., & Goldin, G. A. (2008). Students' expression of affect in an inner-city simcalc classroom. *Educational Studies in Mathematics, 68,* 131–148.

Schutz, P. A., & DeCuir, J. T. (2010). Inquiry on emotions in education. *Educational Psychologist, 37*(2), 125–134.

Sewell, K. (2011). Researching sensitive issues: A critical appraisal of 'draw-and-write' as a data collection technique in eliciting children's perceptions. *International Journal of Research and Method in Education, 34*(2), 175–191.

Skemp, R. (1977). *The psychology of learning mathematics.* Harmondsworth, England: Pelican Books.

Skemp, R. (1979). *Intelligence, learning, and action.* Chichester, England: Wiley.

Skinner, E., Furrer, C., Marchand, G., & Kindermann, T. (2008). Engagement and disaffection in the classroom: Part of a larger motivational dynamic? *Journal of Educational Psychology, 100*(4), 765–781.

Skovsmose, O. (2005). Foregrounds and politics of learning obstacles. *For the Learning of Mathematics, 25*(1), 4–10.

Smith, A. (2004). *Making mathematics count : The report of professor Adrian Smith's inquiry into post-14 mathematics education.* London, England: The Stationary Office.

Smith, J. A., & Osborn, M. (2003). Interpretative phenomenological analysis. In J. A. Smith (Ed.), *Qualitative psychology: A practical guide to research methods.* London, England: Sage.

Solomon, Y., & Rogers, C. (2001). Motivational patterns in disaffected school students: Insights from pupil referral unit clients. *British Educational Research Journal, 27*(3), 331–344.

Svebak, S. (1993). The development of the tension and effort stress inventory (TESI). In J. H. Kerr, S. J. Murgatroyd, M. J. Apter, J. H. Kerr, S. J. Murgatroyd, & M. J. Apter (Eds.), *Advances in reversal theory* (pp. 189–204). Lisse, The Netherlands: Swets & Zeitlinger Publishers.

The Royal Society. (2008). *State of the nation: Science and mathematics education, 14–19 year olds in science and mathematics in the UK.* London, England: The Royal Society.

Thomas, D. M., & Watson, R. T. (2002). Q-sorting and MIS research: A primer. *Communications of the Association for Information Systems, 8,* 141–156.

Thomas, G. (2009). *How to do your research project.* Thousand Oaks, CA: Sage.

Troman, G., Jeffrey, R., & Raggi, A. (2007). Creativity and performativity policies in primary school cultures. *Journal of Educational Policy, 5,* 549–572.

Turner, J. C., Meyer, D. K., & Schweinle, A. (2003). The importance of emotion in theories of motivation: Empirical, methodological, and theoretical considerations from a goal theory perspective. *International Journal of Educational Research, 39,* 375–393.

Vorderman, C., Budd, C., Dunne, R., Hart, M., & Porkess, R. (2011). *A world-class mathematics education for all our young people.* London, England: Government task force.

Vygotsky, L. S. (1986). *Thought and language.* Cambridge, MA: MIT Press.

Walkerdine, V. (1988). *The mastery of reason.* London, England: Routledge.

Watson, A., & DeGeest, E. (2005). Principled teaching for deep progress: Improving mathematical learning beyond methods and materials. *Educational Studies in Mathematics, 58*(2), 209–234.

Weiner, B. (1990). History of motivational research in education. *Journal of Educational Psychology, 82*(4), 616–622.

Williams, S. R., & Ivey, K. M. C. (2001). Affective assessment and mathematics classroom engagement: A case study. *Educational Studies in Mathematics, 47*(1), 75–100.

Zan, R., & Di Martino, P. (2007). Attitude towards mathematics: Overcoming the positive/negative dichotomy. *The Montana Mathematics Enthusiast, 3,* 157–168.

Zan, R., Brown, L., Evans, J., & Hannula, M. (2006). *Affect in mathematics education: An introduction. Educational Studies in Mathematics, 63*(2), 113–121.

Lightning Source UK Ltd.
Milton Keynes UK
UKOW06f1011280316

271010UK00008B/336/P